Interlinking of Rivers in India

Published by Academic Foundation
in collaboration with
THE NATIONAL CIVIL SOCIETY COMMITTEE
ON INTERLINKING OF RIVERS IN INDIA
(NCSCILR)

The Editors:

Yoginder K. Alagh is currently Chancellor, Nagaland University and President of Human Development Institute. Prof. Alagh was former Minister of Power and Planning and Science and Technology, Government of India. Earlier he was Vice Chancellor of the Jawaharlal Nehru University (JNU), New Delhi.

Ganesh Pangare is a water management expert, currently the Technical Director of the World Water Institute, based in Pune. He has authored several publications, the most recent being *Springs of Life: India's Water Resources,* (Academic Foundation, 2006).

Biksham Gujja is policy advisor of Living Waters Programme at WWF-International. He is also special project scientist based at ICRISAT, Hyderabad. Dr. Gujja is a water policy specialist and associated with several global initiatives.

Interlinking of Rivers in India

Overview and Ken-Betwa Link

Editors:

Yoginder K Alagh • Ganesh Pangare • Biksham Gujja

Academic Foundation

NEW DELHI

Published in 2006 by
ACADEMIC FOUNDATION
4772-73 / 23 Bharat Ram Road, (23 Ansari Road),
Darya Ganj, New Delhi - 110 002 (India).
Phones : 23245001 / 02 / 03 / 04.
Fax : +91-11-23245005.
E-mail : academic@vsnl.com
www.academicfoundation.com

Published in collaboration with
THE NATIONAL CIVIL SOCIETY COMMITTEE
ON INTERLINKING OF RIVERS IN INDIA
(NCSCILR)

Interlinking of Rivers in India: Overview and Ken-Betwa Link
Editors: Yoginder K. Alagh, Ganesh Pangare and Biksham Gujja

ISBN 81-7188-515-2

Printed and bound in India

CONTENTS

List of Figures and Tables

FIGURES

TABLES

Contributors

Yoginder K. Alagh is currently Chancellor, Nagaland University and President of Human Development Institute. Dr. Alagh was former Minister of Power and Planning, and Science and Technology, Government of India. Earlier he was Vice Chancellor of the Jawaharlal Nehru University, New Delhi.

Ganesh Pangare is a water management expert, currently the Technical Director of the World Water Institute, based in Pune. He has authored several publications, the most recent being Springs of Life: India's Water Resources.

Biksham Gujja is policy advisor of Living Waters Programme at WWF-International. He is also special project scientist based at ICRISAT, Hyderabad. He is a water policy specialist and associated with several global initiatives.

Ashok Khosla is President of Development Alternatives, New Delhi. Dr. Khosla has won the United Nations Sasakawa Environment Prize 2002.

Jayanta Bandyopadhyay is an expert in natural resources management and holds a Ph.D in engineering from IIT Kanpur. Currently he is a Faculty Member at the Indian Institute of Management, Kolkata.

Shama Perveen works with Prof. Jayanta Bandyopadhyay at the Indian Institute of Management, Kolkata.

Ramaswamy R. Iyer is former Secretary, Water Resources, Government of India. Until recently he was Research Professor (now Honorary Research Professor) at the Centre for Policy Research, New Delhi.

Hajara Shaik is a sociologist and works as an independent consultant. She is based in Gland, Switzerland.

Kanchan Chopra is currently Director and Professor at the Institute of Economic Growth, Delhi. Biodiversity and its economic valuation, water resource management and common property resources have been the focus of her research in the last two decades.

Himanshu Thakkar works with South Asia Network on Dams, Rivers and People (SANDRP), New Delhi.

Bipin Chandra Chaturvedi works with South Asia Network on Dams, Rivers and People (SANDRP), New Delhi.

Medha Patkar is the founder and senior activist of Narmada Bachao Andolan (Save Narmada Movement) since 1985. She is also the National Coordinator of National Alliance of People's Movements. She was also a member of the World Commission of Dams.

Payal Parekh works with the Narmada Bachao Andolan.

A.D. Mohile retired as the Chairman of the Central Water Commission, Government of India and the Director General of National Water Development Agency. Currently he is a consultant based in New Delhi.

Foreword

In 2002 the Supreme Court issued a directive to the Government of India to complete the Interlinking of Rivers (ILR) project in the next 10 years. Since then, ILR has raised unprecedented expectations, concerns and debate.

ILR is one of the most ambitious projects ever proposed to address the issues of water scarcity and conflicts in India. The backbone of this water infrastructure project, envisages various links between 37 major rivers in the country and is proposed in two parts: the Himalayan links and the Peninsular links. The purpose is to find a permanent solution to droughts and floods, to generate additional water for agriculture, and also to increase India's hydropower capacity.

Some Basic Questions

Two years after the issue gained importance, the status of ILR is still shrouded in uncertainty and ambiguity. One thing, however, seems very clear, its cost. Estimated to cost Rs. 560,000 crore (about $120 billion), this project, if undertaken, would be the single largest in any sector, anywhere in the world.

Obviously, talk of such large sums automatically ensures the support of interest groups in favour of the project. On the other hand, the magnitude of the scheme and the prescribed time frame for its implementation, have raised alarms from concerned groups and experts. The NGO community has repeatedly asked for an informed National Dialogue on ILR before work is actually commenced. Such a dialogue could save money and lives, and avoid conflicts between states. This, however, is possible only with reliable and credible information. The public has the right to access full and detailed analyses of the project and know its ramifications. As of now, the Government has not made such information public – there has been no analysis tabled; nor have there been any pre-feasibility reports, detailed project reports or other studies. The reports of the earlier Task Force and its specific recommendations, if any, have not been made public either. While the NGO community is aware of the sensitivities involving hydrological information related to certain shared rivers, there should be no difficulty in releasing similar information on peninsular rivers. There are other types of information that can be made public without jeopardising any national interest. Lack of information is only serving to creating confusion, conjecture and polarisation of views.

Researchers in India have attempted to analyse this project based on the limited information available, and have expressed *a priori* concerns, which broadly fall into six categories.

Economic Issues

Is this proposed project affordable and feasible? This question may have several components such as:

a) Where is the money going to come from?

b) What are the costs involved?

c) What are the benefits?

d) Who is going to pay operation and maintenance costs?

e) What are the risks?

Legal and Policy Issues

Are there any legal hurdles to the proposed project? Obviously this project is going to transfer water from one state to another; it must take into account that water is a subject that falls under state jurisdiction. There are already several long-standing disputes over water-sharing between states, which have and continue to cost the country a great deal. Before investing such a large sum of money on ILR, we will need to examine very closely the legal issues related to large water transfers from one basin to another, across several states.

Social Issues

How many people will be affected by the proposed project and how? Experience shows us that the issues of displacement and relocation cannot be taken lightly in the Indian context. Obviously, a project of this size will involve large scale displacements and rehabilitation issues. Those people who will inevitably lose land, houses and property will need to be satisfied with what they are going to receive as compensation both in the short term and in the long term. In the light of former experiences, social issues and struggles might well delay the project and increase costs, thereby making this whole project uneconomical. It is, therefore, crucial that social losses be thoroughly assessed before the project is begun.

Technical Issues

Is the proposed project technically feasible at the estimated level of investment? Obviously, this will be taken care of by the technical experts in charge of preparing the detailed project proposals, but it is in our best interests to make this information public so as to get other critical views. Issues to be considered in this context include: availability of water, energy requirements, losses due to evaporation, seepage, etc.

Environmental Issues

What are the environmental and ecological costs of the project and how are they going to be addressed? A project of this magnitude will have major implications on terrestrial, freshwater and coastal ecosystems – and will affect the local micro climate over large parts of the country. Some of these will be beneficial and others will not be. These effects will also have direct economic impact on millions of poor people who are directly dependent on those systems.

Connecting rivers may help balance the water availability among regions but it also means spreading pollution. Issues such as health impacts and soil degradation will have to be taken into account. ILR will directly and often adversely impact many protected areas. The issues of ecological impact will need to be assessed very thoroughly to estimate the cost of such impact or the cost of mitigation measures and add this to the project's total cost. To reiterate, these issues need to be addressed before the project is started, not after.

Strategic Issues

Are there any better alternatives (cheaper, socially-acceptable and ecologically friendly) that meet the same objectives as the proposed project? It is important to closely inspect the rationale for the project itself. Also, issues related to transboundary rivers need to be examined and resolved before the project gets underway.

These issues have been raised by the Civil Society from the outset. So far, there has been no informed debate on these matters. Lack of information not only hampers intelligent debate but also leads to speculation and allegations. Since the Government has not provided even pre-feasibility reports, many individuals and groups have naturally opposed ILR. The Government can only stand to gain by sharing information with the Civil Society. It would help them get feedback, identify critical gaps in information and could bring to light some unpredicted dimensions of social and environmental issues, all of which could be dealt with right at the beginning.

It is not enough for the government to declare that these aspects have been taken care of. Such an approach is not only unacceptable, but might well lead to difficulties and obstacles later when the project is being implemented. If ILR should turn out to be the only possible option to address the water crisis in the country, or the best one, the Government needs to show the public why.

Conclusion

It is impossible to overstate the immense significance of the proposal to interlink the rivers of this country. It will have tremendous impact in many spheres of our nation's future. We will be changing the lives of people not merely in our generation but perhaps for centuries to come. The project will forever alter the physical features of our land; its impact on our ecology and ecosystems will be profound and can never be fully predicted. The project cannot be undertaken without understanding its implications or without the informed consent of the people it affects.

The ramifications of this project are too great for us to continue to avoid debate on it. Open and transparent deliberations are a crucial part of the process; what is more, this must be conducted not merely as a public relations exercise, not merely as public meetings for the sake of form, but as a genuine, credible process where people from all sides of the issue sit together and conduct a decision-making dialogue.

The present book is an attempt by the National Civil Society Committee on Interlinking of Rivers in India to throw open to the public some of the issues related to the Interlinking of Rivers Programme and more so on the proposed Ken-Betwa Link.

Dr. Ashok Khosla

President,
Development Alternatives,
New Delhi

Preface

As India's population continues to rise and opportunities for water use diversify along with the peoples' changing needs, economic opportunities and lifestyles, the problems caused by the country's spatial variation in water availability are becoming ever more acute. While on the one hand, parts of the country are among the wettest on earth or are drained by some of the world's largest rivers, others face dire water deficits made worse by climate change, degraded watersheds and excessive, unregulated withdrawals. It was only a matter of time therefore, before grandiose schemes were proposed to convey surpluses from one location to make up shortfalls in another — a national water grid as it were. In fact, the first such proposals were mooted almost fifty years ago before being fleshed out in one form or another during the 1960s and 70s.

Rapid increases in demand for water in recent years have focused attention on these proposals once again to the extent that suddenly, pressure emanating from prominent political parties, individual political figures, engineering groups, lawyers, and well fomented public interest has resulted in a Supreme Court direction demanding that the Government of India prepares and completes a project to interlink certain of the country's major rivers by the year 2012.

A Task Force was established to undertake the programme which now aims to link all major rivers of the country through about 260 links at the estimated cost of Rs. 560,000 crore. The idea is to control floods and provide water for irrigation and other purposes to drought-prone areas through transfer of water from river basins with 'excess' water to deficient basins. The National Water Development Agency (NWDA), established in 1982 to study the development of Himalayan and peninsular rivers has been working on developing a feasible proposal for inter-basin transfer of water through the interlinking of rivers.

As a vibrant democracy, India has a long way to go in engaging the civil society in the decision making process, in relation to water resource development projects. The Interlinking of Rivers (ILR) project is on the national agenda today. A complex project like this needs to be open for public debate not just as a part of good governance, but also in order to avoid social and ecological costs and to also find better options to meet the same goal. However, the process of developing the project by far has not been very participatory till now. Concerns were raised by the National Civil Society Committee on Interlinking of Rivers in India (NCSCILR) on the Government's mode of generating consensus on the project, without including civil society groups, a deviation from the basic diktat of the Common Minimum Programme (CMP). The CMP mentions that the Government will make a comprehensive assessment of the feasibility of linking the rivers of the country in a fully consultative manner. Not only has this not taken place, but in the absence of adequate information and without understanding some very basic issues, it is not possible to even begin to comment on this massive programme. Some of the important issues are those related to the economic viability of the programme, the environmental costs, the social costs, resettlement and rehabilitation issues, operation and maintenance costs, and opportunities for conflicts such as those related to displacement, sharing of benefits and costs between states, management of river basins shared with neighbouring countries, etc.

Where the Ken-Betwa link which is the priority link of the Government is concerned, for example, several groups active in the local areas have expressed concerns about the water balance studies, and the environmental and social impacts of the link. However, neither their work nor their views have been yet incorporated in the Government's consensus building process. It was only after a lot of efforts made by various groups that the feasibility report of the link has been put up in the public domain. Now, efforts are ongoing to review the various aspects related to the link.

The NCSCILR is also presently reviewing the Ken-Betwa link. At the 7th meeting of the Committee at Orcha, on the Betwa river in Madhya Pradesh, a thorough assessment of the feasibility report prepared by the NWDA was undertaken. The outcome of this is the present publication.

We have divided the book into two Sections. The first section will give the reader an overview of the various aspects of the interlinking of rivers programme in India. The second section is an analysis of the feasibility report of the Ken-Betwa link. The report has been analysed by experts from various angles. The pros and cons have been discussed and an objective and rational view by the various researchers has been put forward.

— Editors

Section 1

Overview

1

A Scrutiny of the Justifications for the Proposed Inter-Linking of Rivers in India

JAYANTA BANDYOPADHYAY
SHAMA PERVEEN

The Background of Inter-basin Water Transfers

In the past few centuries there has been a great escalation in the global freshwater requirements while the per capita availability of it has declined drastically. Search for ways to counteract the challenge of freshwater scarcities had been taken up by communities, national governments and international organisations. Heavy investments in engineering structures have opened the possibilities of withdrawing more and more water from the natural sources like lakes, rivers and the groundwater aquifers. More than half of all accessible global freshwater runoff is currently withdrawn for human uses. The consequence is that drastic reductions have taken place in the amount of water remaining instream causing degradation of the aquatic ecosystems and affecting negatively the various services provided by the ecosystems. The example of the Nile in Egypt, the Ganges in South Asia, the Amu Darya and Syr Darya in Central Asia, the Yellow river in China, the Colorado river in North America, etc. are among the major watercourses whose flows have been obstructed and diverted - to such an extent that for parts of the year, little or none of their freshwater flow reaches the sea (Postel, 2000).

Transfer of water from one river basin to another has been practiced as an exemplary engineering response for meeting the growing water requirements. Water resource engineering has traditionally been fixed on providing supply augmentations as the solutions, ignoring the 'soft' demand-side management options (Biswas, 1979). This has resulted in plans for several inter-regional water transfers (IWT) in many parts of the world. One of the earliest examples of IWT is found in Egypt where, in the Pharaonic era, engineering efforts had been made to moderate the flow of the river Nile. King Mina, ruler of Egypt in the First Dynasty, had constructed large number of canals and bridges to carry the Nile water to lower lands (Abu-Zeid, 1983). Another primordial example of IWT can be given from Japan, where water transfers have been in practice for over several hundred years. Though initially started on a small scale, primarily to serve the purpose of rice cultivation, the expansion and development of cities and industries in recent decades (especially after the World War II), provided the conditions for Japan to take up large-scale IWT projects (Greer, 1983).

Several proposals for major inter-basin water transfers were made in the decade of the 1960s. A new generation of plans was put forward in North America in 1964, headlined by the much talked-about and grandiose North American Water and Power Alliance (NAWAPA) scheme. It included numerous plans for the distribution of water from areas with high precipitation in the North-Western part of North America to less water endowed areas of Canada, United States and Mexico (Biswas, 1978). The immensity of the plan stirred the imagination of many engineers and economists and within five years of NAWAPA being proposed, a whole series of IWT schemes were put forward for the redistribution of the waters of North America (Golubev and Biswas, 1978). Similarly, in the erstwhile Soviet Union new engineering proposals were put forward for large scale transfer of water from the more humid to the less humid regions of the continent (Micklin, 1977; Soviet Geography, 1972). Prior to the 1960s, the Soviet Union depended mostly on traditional irrigation techniques. The 'integrated method of development of desert lands' (Gujja and Shaik, 2004) exemplify the immediate realisations of engineering capability created by the alterations of the natural flows in the rivers on a very large scale on economic growth. Golubev and Vasiliev (1978) have pointed out that, "Inter-regional water transfers are appealing because of the great amount of waters produced, which drastically changes the water situation."

The same concepts of engineering generated the proposal for the South to North Water Transfer Project in China. The Yangtze basin, well-endowed with water, as well as other southern river basins benefiting from the ample summer monsoon precipitations, contribute about 80 per cent of the total annual run-off in China.

However, the water from these rivers are available to only 40 per cent of the arable land. On the other hand, the drier regions in the North and North-East of China, the Huang he (Yellow), the Huai he, the Hai he basins and the Northwest inland region together have more than half of the geographic area with 45 per cent of the arable land and nearly 36 per cent of the population. They, however, have direct access to only 12 per cent of the water resources of the country (Gujja and Shaik, 2004). The Chinese Academy of Sciences conducted field investigations of the water transfer in the upper reaches of the Chang Jiang (Yangtze). When fully developed, the scheme has been proposed to divert 40-50 cu km of water per year from the Yangtze basin to the North China plain, alleviating water scarcity for 300–325 million people living in what even then would be a highly water-stressed region (Berkoff, 2003). In India too, projects such as the Periyar-Vaigai system, Indira Gandhi Canal and Telegu Ganga stand as classic examples of IWT. In the 1970s, Rao (1975) proposed the Ganga-Cauvery link canal that still guides the imagination of the governmental engineers in India, though this proposal did not find support for its being followed up.

Opposition to Large Inter-basin Transfers

Water requirements have traditionally been assessed by planners and developers on the basis of projections of population growth, irrigational needs, industrial production and the change in consumption patterns based on economic development. Engineering designs for projects were accordingly made to provide the appropriate supplies. However, over the past few decades, this traditional approach of engineering has been jolted and perceptible changes are taking place in the way water systems are being perceived or managed. With some of the projects undertaken earlier have started to be seen as sources of further problems, water systems management is seen to be facing a crisis. Need for a fundamental shift away from the present reductionist engineering paradigm to a holistic and inter-disciplinary one has been recognised by many water professionals. As Wolff and Gleick (2002:1) have noted, "The world is in the midst of a major transition in the way we think about – and manage - our vital and limited freshwater resources." Such a statement exemplifies fundamental changes that are taking place in water systems management (Postel, 1997; Reddy, 2002; Seckler, 1996).

As a result of such changes, a more cautious approach towards the design of the IWTs was observed during the 1970s. Subsequently, it led to the rethinking on some of the earlier plans for water transfers. For example, in North America itself, implementation of several projects have been abandoned, modified or at least slowed down. The original proposal for the Texas Water System made in 1968, consisting of a large diversion from the Mississippi river into the state of Texas, is a clear indication of this new trend. The earlier plan for the Texas Water System has been modified so much that it now serves as a negative example of IWTs and inter-basin transfers. The overwhelming attention was focussed on narrow aspects of engineering and economics, while scant perfunctory attention was accorded to the associated ecological-economic aspects (Greer, 1983).

What were the reasons for such a change in the professional view of IWT projects? Firstly, there was strong opposition to the transfer from the basins from which water is being taken out. Secondly, the economic feasibility of many such large transfers was not established in a convincing manner. Thirdly, and most importantly, the cumulative environmental impacts, that got little attention in the initial assessment of the feasibility of the projects, over time started to be conspicuous by their impacts. In short, the decade of 1970s was also the time when, "New approaches to complex river development were according greater recognition to its environmental limits and consequences" (White, 1977).

Availability of extra amounts of water in water scarce areas generates great economic values. For the agrarian societies, water provides the road to prosperity. However, there are also negative economic, social and environmental impacts of large water projects. In the past, these have not been observed so conspicuously. With the passage of time, however, such factors start to become conspicuous and the doubts over their justifications and feasibility have led to intense policy debates. Shao and Wang (2003) suggest that, "Inter-basin water transfer projects are prone to problems and controversies, and may challenge the established basin management, legal system and policy making procedure which are taken for granted until such projects are put under consideration." Howe and Easter (1971) additionally caution that IWT projects are likely to prove expensive to nations, except under certain "rescue operation" type of cases (to replenish the groundwater, for instance). On the other hand, Wells (1971) pointing out the case of Texas Water Systems, notes that water imports to the high plains of Texas are not only economically feasible but also that the state simply cannot afford not to import water to that area. Such statements clearly point to the need for a comprehensive feasibility study and options assessment as decision support for large water projects.

The interconnections between the environmental and economic systems become more evident in the background of a widely acknowledged discernible shift in water resource sector from supply to demand side management (ADB, 1999). Hashimoto *et al.* (1982) have identified three criteria for evaluating the performance of water resources systems, *viz.*, how likely a system is to fail (reliability), how quickly it recovers from failure (resilience) and how severe are the consequences of failure (vulnerability). It is suggested that these criteria assist in the evaluation and selection of alternative design and operating policies for a wide variety of water resource projects (Jain, Reddy and Chaube, 2005). The present period is a very dynamic one for conceptual development in water resource management, as a result the older ideas about IWT are getting replaced by newer ones. A great deal of new work is needed to establish a new paradigm and provide a more penetrating and overarching policy framework. The interlinking of rivers in India is being proposed at this important juncture when a shift in the paradigm of water resource management is taking place. It is stressed here that with the indications of a shift taking place in the paradigm for the management of water systems the world over, proposals for such heavy investments need to be assessed in a very comprehensive and participatory manner.

The Proposed Interlinking of Rivers in India

The distribution of precipitation in India is characterised by wide spatial and temporal variations. This is a direct result of the domination of the Monsoon in the making of the weather of South Asia. While some parts of the region receive very large monsoon precipitation and rivers inundate wide areas in the floodplains, some other parts receive much less precipitation and often face scarcity of water. On the basis of the public statements of the political leaders, the proposed Interlinking of Rivers (ILR) may be justified as the win-win solution to the dual problems of floods and droughts in the various parts of the country. In view of the fact that the proposal is billed as the largest engineering project in the world, such a proposal, which appears to offer a simple solution to a complex problem, needs to be assessed with a high level of transparency and professionalism. This becomes particularly important, and difficult, because almost all technical information on the proposed ILR has not been put in the public domain. This acts as a great obstacle to the growth of better scientific knowledge on the problems and taking of appropriate decisions to address them.

India occupies about 2.45 per cent of the terrestrial surface of the Earth and 72 per cent of the geographical area of South Asia. When viewed in terms of precipitation per unit area of the land, the country receiving 4 per cent of the total global precipitation, seems to be well above the global average. Nonetheless, with over 17 per cent of the world's population living in the country, its position in terms of per capita water availability is seen to be quite difficult. The spatial and temporal variations in the precipitation lead to regional inequities in water availability in India (Figure 1.1). About 71 per cent of the available water resources of India is localised in 36 per cent of the geographical area of the country, primarily in the Ganges-Brahmaputra-Meghna basin and all the west flowing rivers from the Western Ghats.

Such spatial and temporal variations often give rise to conditions of water scarcity existing in some parts, while rivers inundating large areas in some others. Following the priority given to irrigation in the official policy, and the availability of the green revolution packages, the demand on the country's water resources from the irrigation sector has substantially grown. Irrigation potential has grown from about 22 Mha during the early 1950s to about 105 Mha by the turn of the century. For both agriculture and industry, every additional volume of water opens up scope for increased financial return. In many parts of the country, the greater purchasing power of the industry has caused either over-exploitation of groundwater resources or diversion of water from sources traditionally used for irrigation. These conflicting demands for transfer of water from various regions for the provisioning of domestic supplies, for promoting commercial agriculture or industries have started to show the signs of becoming potential basis for intense conflicts over the available resources.

Figure 1.1

Spatial Variation in Precipitation Pattern over the Country

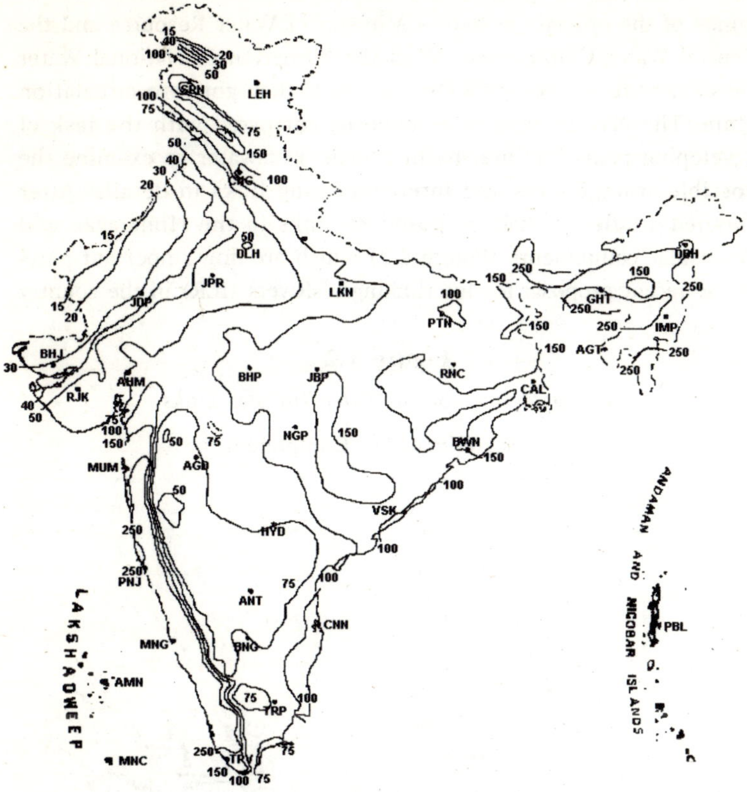

Source: http://www.imd.ernet.in/section/climate/annual-rainfall.htm

About a century ago, mainly for improving navigational access, proposals for several interlinking canals were made by Arthur Cotton. With the availability of greater engineering capabilities, the idea of transferring large quantities of water from river basins with higher rainfall to the drier basins with the objective of irrigation started to encourage policymakers to think on IWTs. In this line, Rao (1975) proposed three link canals between Brahmaputra and Ganga, between Ganga and Cauvery, and between Narmada and parts of Rajasthan with the purpose of transferring water to the drier areas in southern and western India. Such a proposal was not favoured on various grounds by the Central Water Commission of India, including the cost.

The main idea of transferring water from the Ganga-Brahmaputra river system to the less water endowed areas in southern and western parts of India by linking canals, nevertheless, remained alive in the minds of the officials in India's Ministry of Water Resource and the Central Water Commission. With the formation of National Water Development Agency (NWDA) in 1982, they got into circulation again. The NWDA was, subsequently, entrusted with the task of developing plans for inter-basin transfers of water to examine the possible storage sites and interconnecting links in details. After detailed studies, it had proposed 30 links in the Himalayan and Peninsular components (Figure 1.2) which are now important parts of the recent proposal for Interlinking of Rivers (ILR) in the country

Figure 1.2

Proposed Inter-Basin Water Transfer Links

1.2(a) Himalayan Component

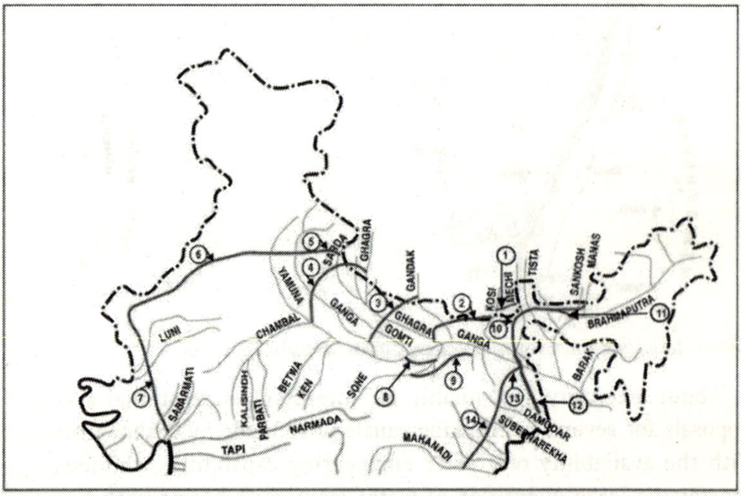

1. Kosi — Mechi
2. Kosi — Ghagra
3. Gandak — Ganga
4. Ghagra — Yamuna*
5. Sarda — Yamuna*
6. Yamuna — Rajasthan
7. Rajasthan — Sabarmati
8. Chunar — Sone Barrage
9. Sone Dam — Southern Tributaries of Ganga
10. Manas — Sankosh - Tista — Ganga
11. Jogighopa — Tista — Farakka (Alternate)
12. Farakka — Sunderbans
13. Ganga (Farakka) — Damodar — Subernarekha
14. Subernarekha — Mahanadi

*FR Completed

1.2(b) Peninsular Components

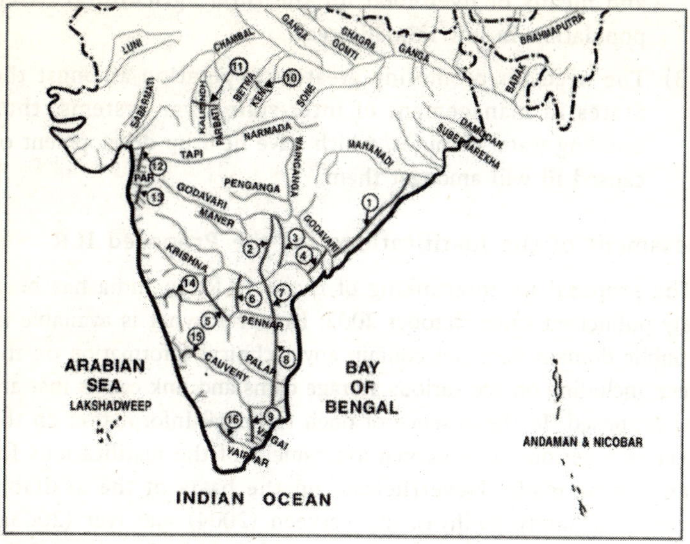

1. Mahanadi (Mahabhadra) — Godavari (Dowlaiswaram)*
2. Godavari (Inchampalli) — Krishna (Nagarjunasagar)*
3. Godavari (Inchampalli) — Krishna (Pulichintala)*
4. Godavari (Polavaram) — Krishna (Vijaywada)*
5. Krishna (Almatti) — Pennar*
6. Krishna (Srisailam) — Pennar*
7. Krishna (Nagarjunasagar) — Pennar (Somasila)*
8. Pennar (Somasila) — Palar-Cauvery (Grand Anicut)*
9. Cauvery (Kattalai) — Vaigai — Gundar*
10. Ken — Betwa*
11. Parbati — Kalisindh — Chambal*
12. Par — Tapi — Narmada*
13. Damanganga — Pinjal*
14. Bedti — Varda
15. Netravati — Hemavati
16. Pamba — Achankovil — Vaippar*

*FR Completed

(TFILR, 2003). The justification presented for the proposed ILR by the TFILR (2003:22) is based on addressing three types of need:

1) The paramount need for national self-sufficiency in food and energy with sustainability.

2) The need for regional equity in regard to poverty alleviation and means of livelihood for the rural agricultural based population in low rainfall areas.
3) The need for promoting greater cooperation amongst the States in management of inter-state river systems, thus avoiding water disputes, which have held up development or caused ill will amongst them.

Assessment of the Justifications for the Proposed ILR

The proposal for interlinking of rivers (ILR) in India has been widely publicised since October 2002. However, what is available in the public domain does not contain any technical information on the project, including on the various storage dams and link canals that are being proposed. In the absence of open technical information on the project, any serious professional assessment of the justifications for it can not be made. Nevertheless, on the basis of the available information, Bandyopadhyay and Perveen (2004) and Iyer (2003a) have drawn attention to why an open professional review of the proposed ILR is very much needed. After years of repeated public demand from many professional quarters to the Ministry of Water Resources, some details of only one link, the Ken-Betwa link, has recently been put up in the public domain. The National Civil Society Committee on Interlinking of Rivers (NCSCILR) has arranged for a review of the Feasibility Report for this proposed link, which is presented later in the book. This amply substantiates the expressed need for opening the assessment to non-governmental professionals. In addition, significant questions have been raised by many on the conceptual framework, justifications and options assessment, as given in favour of the proposal (Iyer, 2004, Vaidyanathan, 2003). Proper study of these critical comments and transparency in addressing them is of utmost importance for taking the correct decisions regarding the ILR.

Conceptual Gaps in the Proposed ILR

The main conceptual framework that is used to design the proposed ILR is based on an identification of the various river basins of India under broad categories of surplus, marginally surplus, marginally deficit and deficit. Though the World Water Assessment

Berkoff, J. (2003) "China: The South-North Water Transfer Project - Is It Justified?' *Water Policy* 5(1):1-28.

Bhushan, P. (2003) *The River Linkages Payoff* available online (http://indiatogether.org/2003/jun/opi-rivers.htm)

Biswas, A.K. (1978) "North American Water Transfers: An Overview" *Interegional Water Transfers: Problems and Prospects Vol. 6* (Oxford: Pergamon Press).

Biswas, M.R. (1979) *Agriculture and Environment Technical Memoir* No. 3, pp. 225-255, International Commission on Irrigation and Drainage, New Delhi.

Blaikie, P.M. and J.S.S. Muldavin "Upstream, Downstream, China, India: The Politics of Environment in the Himalayan Region" *Annals of the Americal Association of Geographers 94*(3):520-548.

Bruijnzeel, L.A. (2004) "Hydrological Function of Tropical Forests: Not Seeing the Soil for the Trees?" *Agriculture, Ecosystem and Environment* 104:185-228.

Carruthers, I. and J. Morrison (1994) "2020 Vision - Dramatic changes in the world Agricultural and Industrial Production Systems" *IIMI Review*, 8 (1):14 - 20.

Chalise, S.R. and N.R. Khanal (1996) *Extended Abstracts: International Conference on Ecohydrology of High Mountain Areas* (Kathmandu:ICIMOD).

Chitale, M.A. (1995) "Institutional Characteristics for International Cooperation in Water Resources" in *Water Resources Development* 11(2):113-123.

Cosgrove, W. and F.R. Rijsberman (2000) *World Water Vision: Making Water Everybody's Business*, Earthscan, London.

Dunbar M.J and M. C. Acreman (2001) 'Applied Hydro-Ecological Science for the Twenty-First Century' in M.C. Acreman (Ed.) *Hydro-cology: Linking Hydrology and Aquatic Ecology IAHS* No. 266, 162 pp, Centre for Ecology and Hydrology, IAHS Press, Wallingford.

Falkenmark, M.; J. Lundqvist and C. Widstrand (1989) "Macro-Scale Water Scarcity Requires Micro-Scale Approaches: Aspects of Vulnerability in Semi-Arid Development" *Natural Resources Forum* Nov: 258-267.

Falkenmark, M. and J. Lundqvist, (1995) "Looming Water Crisis: New Approaches are Inevitable" in L. Ohlsson (Ed) *Hydropolitics*, pp. 178 -212, Zed Books, London.

Feitelson, E. and J. Chenoweth (2002) "Water Poverty: Towards a Meaningful Indicator" *Water Policy*, 4(3):263-281.

Fisher, F.M. (1995) "The Economics of Water Dispute Resolution, Project Evaluation and Management: An Application to the Middle East" in *Water Resources Development* 11(2):377-390.

Gaur, V.K. (ed.) (1993) *Earthquake Hazard and Large Dams in the Himalaya*, Indian National Trust for Art and Cultural Heritage, New Delhi.

GEF (1995) *Scope and Preliminary Operational Strategy for International Waters*, pp. 1-3, Global Environment Facility, Washington DC.

Gleick, P.H. (1999) "The Human Right to Water." *Water Policy*, Vol. 1, No. 5, pp. 487-503.

———. (2000) *The World's Water 2000-2001: The Biennial Report on Freshwater Resources*, Island Press, Washington DC.

Golubev, G. and A.K. Biswas (1978) "Inter-regional Water Transfers" in *Inter-egional Water Transfers: Problems and Prospects* Vol. 6, Pergamon Press, Oxford.

Golubev, G. and O. Vasiliev (1978) "Inter-regional Water Transfers as an Inter-disciplinary Problem" in Biswas, A.K. *et al.*, (Eds) (1983) *Long-Distance Water Transfer - A Chinese Case Study and International Experiences*, UNU, Tycooly.

Greer, R. (1983) "The Texas Water System: Implications for Environmental Assessment in Planning for Interbasin Water Transfers" in Biswas, A.K. *et al.*, (Eds) (1983) *Long-Distance Water Transfer - A Chinese Case Study and International Experiences*, UNU, Tycooly).

Gujja, B. and H. Shaik (2004) "Linking of Rivers: Lessons from the Past" in M. Patkar (Ed.) *River Linking: A Millennium Folly?*, NAPM, Mumbai.

Haddad, Brent M. (2000) *Rivers of Gold: Designing Markets to Allocate Water in California*, Island Press, Washington DC.

Hanchate, A. and T. Dyson (2004) "Prospects for Food Demand and Supply" in Dyson, T., R. Cassen and L. Visaria (Eds.) *Twenty-first Century India: Population, Economy, Human Development and the Environment*, OUP, New Delhi.

Helming, S. and J. Kuylenstierna (2001) "Water–A Key to Sustainable Development" *Issue Paper for the International Conference on Freshwater*, 03-07 December, Bonn.

Homer-Dixon, T. (1994) "Environmental Scarcities and Violent Conflict" *International Security* 19(I):5-40.

Howe, C.W., and Easter K.W. (1971) *Interbasin Transfers of Water: Economic Issues and Impacts*, Johns Hopkins Press, Baltimore.

Hukka, J.J. and T.S. Katko (2003) *Water Privatisation Revisited: Panacea or Pancake?*, IRC International Water and Sanitation Centre Delft, The Netherlands.

Ives, J. D. and B. Messerli (1989) *The Himalayan Dilemma*, Routledge, London.

IWMI (2000) *World Water Supply and Demand:1995 to 2025*, IWMI, Colombo.

Iyer, R.R. (2004) "River-Linking Project: Many Questions" in M. Patkar (Ed) *River Linking*: A Millennium Folly, NAPM, Mumbai.

———. (2003a) 'Linking Rivers...A Chimera of a Project' *The Hindu Survey of the Environment*.

———. (2003b) *Water Perspective, Issues, Concerns*, pp. 368, Sage, New Delhi.

Khattri, K.N. (1987) "Great Earthquake Seismicity Gaps and Potential for Earthquake Disaster along the Himalayan Plate Boundary" *Techno-Physics*, 138:79-82.

Landry, C.J. (1998) "Saving Our Streams Through Water Markets: A Practical Guide" *Political Economy Research Centre*, Bozeman, Montana.

Lipton, M; J. Litchfield and J. M. Faurès (2003) "The Effects of Irrigation on Poverty: A Framework for Analysis" *Water Policy*, 5(5-6):413-427.

Mallik, B. and S. Perveen (2004) "A Scientific and Economic Assessment of the River Link Project in India" in Proceedings of the *Workshop on Interlinking of Rivers: Doable and Desirable?*, West Bengal Academy of Science and Technology, Kolkata.

Martin, B. (1996) "From the Many to the Few. Privatisation and Globalisation" in *The Ecologist* 26(4):145-155.

Micklin, P.P. (1977) "NAWAPA and Two Siberian Water-Diversion Proposals: A Geographical Comparison and Appraisal" *Soviet Geography: Review and Translation*, 18(2):81-99.

Mishra, D.K. (2003) "Life Within the Kosi Embankments" *Water Nepal* 9-10(1-2):277-301.

MoA (2003) *Annual Report 2002-2003*, Department of Agriculture and Cooperation, Ministry of Agriculture, New Delhi, http://agricoop.nic.in/cropprod02.htm

Mohanty, N. and S. Gupta (2002) "Breaking the Gridlock in Water Reforms through Water Markets: International Experience and Implementation Issues for India" *Working Paper Series*, Julian L. Simon Centre for Policy Research.

Mohile, A.D. (1998) *India's Water and its Plausible Balance in Distant Future* Unpublished Paper as Cited in NCIWRDP (1999a:29).

Molden, D. and C. de Fraiture (2004) "Investing in Water for Food, Ecosystems and Livelihood" *Blue Paper*, Stockholm, CGIAR.

Molle, F. and P. Mollinga (2003) "Water Poverty Indicators: Conceptual Problems and Policy Issues," *Water Policy*, 5(5-6):529-44.

NCIWRDP (1999) *Integrated Water Resource Development: A Plan for Action*, National Commission on Integrated Water Resource Development Plan, MoWR, New Delhi.

NCIWRDP (1999a) *Report of the Working Group on Interbasin Transfer of Water,* National Commission on Integrated Water Resource Development Plan, MoWR, New Delhi.

Nigam, A., B. Gujja, J. Bandyopadhyay and R. Talbot (1997) *Freshwater for India's Children and Nature,* WWF and UNICEF, New Delhi.

OED (1995) *Learning from Narmada OED Precis No 88,* The World Bank, Washington DC.

Patkar, M., L. C. Jain, K. Nayar, S.G.Vombatkere, H. Thakkar and R. R. Iyer (2005) *President Kalam, Please Listen,* Available online *http://indiatogether.org/2005/may/env-kalamletr.htm*

Polak (2004) "Water and the Other Three Revolutions Needed to End Rural Poverty." Prepared For the Stockholm Water Symposium.

Postel, S. (2000) "Troubled Waters," *Sciences,* March, New York Academy of Sciences.

———. (1999) *Pillar of Sand: Can the Irrigation Miracle Last?,* WW Norton, New York.

———. (1997) *The Last Oasis: Facing Water Scarcity,* WW Norton, New York.

Rao, K.L. (1975) *India's Water Wealth,* Orient Longman, New Delhi.

Reddy, V. Ratna (2002) "Water Security and Management: Lessons from South Africa" *Economic and Political Weekly* 37(28):2878-81, 13 July.

Rajagopal, A. (2002) *Aquaculture 'Prosperity' – Some Reflections from India,* South Asian Consortium for Interdisciplinary Water Resources Studies, Hyderabad.

Rasinmäki, J. (1997) "The Privatisation of the Administration of Local Government: A Judical Study of the Background, Provisions, Restrictions, forms and Effects of Privatisation." *Kauppakaari Oy.* p. 512 Helsinki.

Raskin, P; P. Gleick; P. Kirshen; G. Pontius and K. Strzepek (1997) "Water Futures: Assessment of Long-range Patterns and Problems," *SEI*) pp 77, Stockholm.

Rath, N. (2003) "Linking Rivers: Some Elementary Arithmatic" *Economic and Political Weekly* 38(29):3033.

Rijsberman, F. (2003) "Can Development of Water Resources Reduce Poverty?" *Water Policy* 5(5-6):399-412.

Rosegrant, M., X. Cai and S. Cline (2002) *World Water and Food to 2025: Dealing with Scarcity,* International Food Policy Research Institute, Washington DC.

Salameh, E. (2000) "Redefining the Water Poverty Index" *Water International* 25(3):469-473.

Saejis, H.L. and M. Van Berkel (1995) "Global Water Crisis: Major Issues of the 21st Century, a Growing and Explosive Problem" *European Water Pollution Control* 5(4):26-40.

Salman, M.A. and K. Uprety (2002) *Conflict and Cooperation on South Asia's International Rivers: A Legal Perspective,* pp. 217, The World Bank, Washington DC.

Seckler, D. (1996) "The New Era of Water Resource Management: From Dry to Wet" *Water Savings Research Reports No. 1,* IIMI, Colombo.

———. (1992) "The Sardar Sarovar Project in India: A Commentary on the Report of the Independent Review" *Discussion Paper 8,* Winrock International, Water Resources and Irrigation Division.

Seckler, D., D. Molden, U. Amarasinghe, C. de Fraiture (2000) "Water Issues for 2025: A Research Perspective" IWMI's Contribution to the *Second World Water Forum.*

Seckler, D; U. Amarasinghe; D. Molden; R. De Silva and R. Barker (1998) *World Water Demand and Supply, 1990 to 2025: Scenarios and Issues,* IIMI, Colombo.

Shah T., M. Alam, D. Kumar, R.K. Nagar and M. Singh (2000) "Pedaling Out of Poverty: Social Impact of a Manual Irrigation Technology in South Asia" *Report No. 45,* IIMI, Colombo.

Shah, T. and O.P. Singh (2003) "Can Irrigation Eradicate Rural Poverty in Gujarat?" *Water Policy Research Highlight 10,* IWMI-TATA Water Policy Program, Anand.

Shao, X. and Wang, Z. (2003) "Interbasin Transfer Projects and their Implications: A China Case Study" in *International Journal of River Basin Management,* 1(1):5-14.

Sharma, S. (2003) "Budgeting Interlinking of Rivers" *Countercurrents*. Org, Online edition *(http://www.countercurrents.org/en-sharma.htm)*

Shiklomanov, I. (2000) "Appraisal and Assessment of World Water Resources" *Water International* 25(1): 11-32.

Singh, B. (2003) "A Big Dream of Little Logic" *The Hindustan Times*, March 09, New Delhi.

Smakthin, V., C. Revenga and P. Döll (2004) "Taking into Account Environmental Water Requirements in Global-scale Water Resources Assessments" *Comprehensive Assessment Research Report No. 2*, International Water Management Institute, Colombo.

Small, L. and I. Carruthers (1993) *Farmer Financed Irrigation: The Economics of Reform*, Cambridge University Press, Cambridge.

Snaddon, C.D., M.J. Wishart and B.R. Davies (1998) "Some Implications of Inter-basin Water Transfer for River Ecosystem Functioning and Water Resources Management in Southern Africa" *Aquatic Ecosystem Health and Management* 1(2):159–182.

Soviet Geography: Review and Translation (1972) Vol 8, November.

Starr, Joyce. (1991) "Water Wars" *Foreign Policy* No. 82.

Swaminathan, M.S. (1999) *A Century of Hope: Harmony with Nature and Freedom from Hunger*, East West Books, Madras.

TFILR (2003) *Interbasin Water Transfer Proposals*, MoWR, New Delhi.

Vaidyanathan, A. (2003) "Interlinking of Rivers" *The Hindu* (Chennai) 26 March.

Valdiya, K.S. (1992) "Must We Have High Dams in the Geo-Dynamically Active Himalayan Domain" *Current Science* 63(6):289-296.

Vance, S., M. Bickle, S. Ivy-Ochs and P.W. Kubik (2003) "Erosion and Exhumation in the Himalayas from Cosmogenic Isotopic Inventories of River Sediments" *Earth Planetary Science Letters*, 206, 273-288.

Verghese, B.G. (2003) "Waiting for Godaganga" *Outlook Magazine* 30 June.

Verghese, B.G. *et al.*, (1994) *Converting Water into Wealth*, Konark, New Delhi.

Vidal, J. (2003) "India's Dream Bangladesh's Disaster" *The Guardian* 24 July *(http://www.countercurrents.org/en-vidal240703.htm)*

Wang Shucheng (2002) *Resource Oriented Water Management*, China WaterPower Press, Beijing.

Wells, D.M. (1971) "High Plains Irrigation and Texas Water Plan" *Journal of Irrigation and Drainage Division*, pp. 123-130, American Society of Civil Engineering,.

Westing, A.H. (Ed.) (1986) *Global Resources and International Conflict: Environmental Factors in Strategic Policy and Action*, Oxford University Press, New York.

White, G. (1977) "Comparative Analysis of Complex River Development" in Gilbert F. White (ed.) *Environmental Effects of Complex River Development*, Westview Press, Boulder.

Wolf, A. (1998) "Conflict and Co-operation along International Waterways" *Water Policy*, 1 (2): 251–265.

Wolff, G. and P. Gleick (2002) "The Soft Path for Water" in Peter Gleick (Ed) *The World's Water: 2002-2003*, Island Press, Washington DC.

World Commission on Dams (WCD) (2000) *Dams and Development – A New Framework for Decision Making*, Earthscan, London.

World Bank (1999) *India: Water Resource Management - The Irrigation Sector*, Allied Publishers, New Delhi.

World Water Assessment Program (WWAP) (2001) "Indicators for the World Water Development Report: From Water-Related Stress to Sustainable Water Use" *Tech Note 1*.

2

River-linking Project: A Critique

RAMASWAMY R. IYER

Background

The previous Government had announced a massive 'Inter-linking of Rivers' Project (ILR) in 2002. While this was welcomed by many, serious doubts and reservations were expressed by others. Subsequently, a change of Government took place, and the Common Minimum Programme of the United Progressive Alliance (UPA) Government stated that a comprehensive assessment of the Project would be made in a fully consultative manner. This has not taken place so far. However, it appears from media reports that the present Government proposes to take up the links individually in a selective manner. It has also set up a Committee of Experts in this context. Meanwhile, however, the Supreme Court continues to monitor the Project, and the Government appears to have assured the Learned Judges that it is proceeding with the Project. Moreover, in various speeches on different occasions, the Hon'ble President of India has also been commending the Project as an answer to the country's problems of flood and drought and future water scarcity. There are grounds for fearing that continued endorsement of the Project by the Hon'ble President and monitoring of the Project by the Supreme Court might make any rigorous and objective examination of the Project in accordance with the usual prescribed procedures extremely difficult, if not impossible.

Two Ambiguities

As we proceed to look at the project, we encounter two basic ambiguities.

One claim, often heard, is that this is a 'mega' project, a major initiative, one of the biggest undertakings in the world, bigger than

China's Three Gorges Project, and so on. There is much professional pride and nationalistic sentiment in such statements. On the other hand, it is also stated that this is only a concept and not a project, and that it will be proceeded with gradually and carefully, link by link. There is either a lack of clarity here, or a deliberate ambiguity.

Similarly, we are sometimes told that much work has been done on the project, that there are pre-feasibility studies relating to almost all the proposed links and that in respect of nine links feasibility reports are available, but on other occasions it is stated that there is no project as yet and that things are at a very preliminary stage.

The following questions arise: Are we talking about a concept or about a project? In the latter case, is it one 'mega' project or 30 individual projects? How many pre-feasibility and feasibility reports are available? Can we see them? Unless these are put in the public domain, how can the CMP's promise of a comprehensive assessment in a fully consultative manner be fulfilled?

'Concept'

Let us look at the 'concept'. It is a rather odd concept, though we may have failed to notice this. One would expect the water planners to start from the identification of the needs of water-scarce areas in Rajasthan, Madhya Pradesh, Maharashtra, Karnataka, Andhra Pradesh, and Tamil Nadu, and work out area-specific answers to their problems. In each case, if we wish to proceed scientifically, the following questions will have to be asked: What is the extent of the problem? To what extent can local and intra-basin answers be found? Is there an inescapable need for bringing in water from outside? If that is so in a given case, and if that necessitates the linking of two rivers as the best answer, one can take a careful look at such a proposal. However, the project *inverts* that logical sequence: it starts with the *a priori* proposition that the rivers of India should be linked, and then proceeds to consider the modalities of doing so.

References are made to a 'national water grid' on the analogy of a power grid or the linking of highways. The analogy is inapt and misleading. In a power-grid or a highway-link, the movement can be in both directions, but that is not the case with a river-link; water will flow only in one direction. Apart from that, highways and power lines are human creations and can be manipulated by humans. Rivers are not human artefacts; they are natural phenomena, integral components of ecological systems, and inextricable parts of the cultural, social, economic, spiritual lives of the communities concerned. They are not pipelines to be cut, turned around, welded and rejoined.

Justifications for Project

Continuing at the 'concept' level, let us take note of the two main justifications offered for the 'inter-linking of rivers.'

The first is that it is an answer to the occurrence of floods in some parts of the country and drought in other parts; that the project will transfer water from the former areas to the latter, providing some relief from floods to the former areas and making more water available to the latter.

The second is that some river basins are 'surplus' in water while others are 'deficit', and that the project will transfer water from the former to the latter.

It must be noted that these are two distinct propositions, but they are often mixed up. We must be clear as to what the basic rationale of the project is.

'Surplus' and 'Deficit'

Taking first the justification of transfers from surplus to deficit basins, there is the crucial question of how 'surpluses' and 'deficits' are determined. The assumptions behind the calculation of a 'surplus' will have to be gone into thoroughly, and even small changes in those assumptions and other parameters may affect the calculation substantially. This has been shown by Prof. A. Vaidyanathan in an article in the Economic and Political Weekly (July 2003). Other scholars, for instance Dr. Jayanta Bandyopadhyay of IIM Calcutta, point out that the very idea of a 'surplus' in a river is problematic. Apart from meeting the needs of drinking water, irrigation, industrial uses, etc., the river also serves ecological and other purposes, so that *any* diversion of waters from it, however small, is bound to have some consequences, major or minor. Similarly, it will be necessary to examine whether a projected 'deficit' is in fact the result of bad water management and unsustainable demands. If it is, the deficit will disappear with better water management. Careful, economical and sustainable intra-basin management should come first, and bringing water from elsewhere should be the last recourse. Further, even in the States that are presumed to be water-rich (*e.g.*, Bihar, Orissa, Andhra Pradesh) there are problem areas, and 'surpluses', if any, should perhaps be first used in those areas rather than sent to distant places.

It is possible that all these aspects have been considered by the NWDA in its pre-feasibility and feasibility reports, but as mentioned earlier, those reports are not available to us.

Flood Diversion

Turning to the other justification, i.e., the diversion of flood waters, the 'paradox of flood and drought' is a formulation often heard in India. There is neither 'paradox' nor 'irony' here: floods and droughts are merely natural phenomena that impinge on our lives, depending on where we live. It is hot and humid in Kolkata and Chennai and cool and pleasant in Darjeeling and Ooty, but no one has found any irony or paradox here or suggested that the temperature variations should be evened out. What we should do about the occurrence of floods in one area, and what we should do about the water needs of arid or drought-prone or water-short areas, are two separate questions. Area-specific ways of coping with these features of nature have to be, and can be, worked out. This is where the challenge lies. Unfortunately, the wrong perception of a paradox or irony here leads to the wrong answer, namely, the 'inter-linking of rivers' to divert flood-waters to arid areas.

The very idea of flood control has been questioned by some, but leaving that aside, it is not quite clear how the linking of rivers will contribute to the objective of flood control. A significant moderation of floods will call for a massive diversion of flood-waters which may not be feasible at all, or if technically feasible, it may have serious impacts on the river regime downstream of the diversion point, on the diversion route, and in the recipient areas. On the other hand, if only small fractions of the flood flows are to be diverted (as seems to be the intention), there will be hardly any flood-moderation. Even if all the river-linking proposals are implemented, the contribution that this will make to the mitigation of the flood problem will not be substantial. Dr. Bharat Singh, Professor Emeritus at the Water Resources Training Centre at the IIT, Rourkee, and Member of the National Commission for Integrated Water Resources Development Plan (1996-99), has observed categorically that "any water resources engineer will immediately discard the idea of the inter-linking of rivers as a flood control measure" (The Hindustan Times, 9 March 2003).

However, if we wish to put aside those doubts and undertake a more detailed examination of this matter, the questions to be considered will be the following: (a) How much of the flood is proposed to be transferred? (b) What degree of flood moderation will this provide to the area in question? (c) What has been the past experience with flood-moderation through big projects? (c) What will be the impact of the diversion on the downstream river regime, aquatic life, livelihoods of riparian communities, water quality and self-purifying capacity, groundwater re-charge, estuarine conditions, etc.? (d) What will be the impacts (good and bad) of the water *en route* as it travels to the destination? (e) What will be the impacts (good and bad) of the transfer of water on the recipient river/area?, and (f) Are the floods in the 'donor' area and the droughts in the recipient area synchronous, and if not, where will the waters be stored temporarily, what will be the impacts of those storage projects? (These questions will also be relevant in the examination of transfers of surpluses to deficit areas.) Once again, no detailed information has been made available.

Power Generation

A further justification offered for the project is that there will be a net generation of electric power (of upwards of 30000 MW). One would normally expect projects for inter-basin transfers to *require* large quantities of energy (for lifting and/or transporting water). How a project of this kind will be a *net generator* of large quantities of electric power is not very clear. If 30000 MW is the net generation, what will be the gross generation, and in which projects will it be generated? Some of that energy will no doubt be consumed by the river-linking processes; hence the idea of 'net generation'. If so, the generation of power is attributable to the large storage projects and not to the linking of rivers (which will actually consume some of that power). This needs careful examination.

Answer to Drought?

It is primarily in the context of drought that the project might appear to be needed. However, two points must be noted. *In the first place, the proposed river links (reportedly mainly by gravity, with a few modest lifts) are no answer to drought-prone areas.* Linking a river to another will merely provide additional water to areas already served by rivers. Most of the uplands and dry lands of this country are distant from rivers, and at elevations of 300 m to 1,000 m above mean sea level. The ILR will serve very few such areas. *The second and complementary point is that fortunately no such long-distance transfer is necessary.* There are several well-known examples of the transformation that can be brought about through local rainwater-harvesting and watershed development even in low-rainfall areas. In brief, the primary answer to drought has to be local; it is only thereafter, and in some very unpromising places where rainwater-harvesting may not be feasible or may yield meagre results, that the bringing in of some external water may need to be considered.

Water for Irrigation?

Turning to water transfers for irrigation, these may be proposed either for providing additional water to areas already under irrigation or for extending irrigation to arid or 'rainfed' areas. In both cases, difficult questions arise.

In irrigated areas, the question is whether large demands for additional irrigation water should be unquestioningly accepted and met through supply-side solutions such as large dams or inter-basin transfers. Should not serious attempts be made to improve water-use efficiency in irrigated agriculture, get more value out of a given quantum of water, reduce the water-demand, and minimise the need for supply-side projects? In the context of the prevailing low efficiency of water-conveyance in canal systems and water-use in irrigated agriculture, bringing in more water from another basin would really amount to the provision of more water for being wasted. It would also mean that there would be no motivation at all for changing cropping patterns and shifting from water-intensive crops to crops that need less water; on the contrary, the tendency to grow water-consuming crops would receive strong encouragement.

In arid or drought-prone areas, the introduction of irrigated agriculture of a kind appropriate to wet areas may be unwise. It can be, and has been, argued that the Rajasthan Canal project was not a good idea but a misconceived one. That may be a controversial statement, but the slogan of 'making the desert bloom' is not necessarily a sound one. 'Development' in arid areas should perhaps take other, less water-intensive forms.

In both irrigated and rainfed areas, the importation of external water may also have other secondary consequences: the increased incidence of conditions of water-logging and salinity (a concomitant of irrigated agriculture in many places); the possibility of the repetition of the 'Green Revolution' patterns of agricultural development and the related problems (which need not be spelt out); and so on. These are not unavoidable consequences of additional water, but they are dangers that have to be kept in mind: given more water, there would be a natural tendency to slip into familiar agricultural practices.

Gigantism/Altering Nature

Apart from the questions raised above about the principal justifications put forward for the project, misgivings arise from the grandiose nature – the gigantism - of the undertaking. This will be a massive intervention in nature, a severe case of technological hubris, of a kind that (one thought) had been discredited and was a thing of the past. Apprehensions about gigantism have been sought to be set at rest with the explanation that the flows will be largely by gravity with lifts (not exceeding 120 metres) at a few selected points, and that the need for a transfer of water through natural barriers will be obviated. Two questions arise. First, such an approach may be possible in some cases, but its feasibility in some 30 projects seems *prima facie* doubtful. This, like the claim referred to earlier that the project will be a net generator of large quantities of electricity, needs to be looked at very carefully, case by case. Secondly, if indeed the links are to be largely by gravity with a few modest lifts, will not such an approach limit the extent of water transfers and the scope of the project? Can large claims still be made for the project?

Impacts and Consequences

The project is potentially fraught with serious consequences. It will necessarily involve dams, reservoirs, diversion of waters, canal systems, and so on. By now there is adequate knowledge of what all this entails, and it need not be spelt out in detail. The adverse impacts and consequences of such large-scale interventions have been observed in many projects, and will need to be studied carefully in the case of each of the proposed links. Leaving aside submergence, displacement, impacts on flora and fauna, and so on, let us take note of one point, namely, that before diverting waters and reducing downstream flows we must make sure that serious consequences will not follow. It is well-known that old-style planning in the former Soviet Union led to the diversion of two rivers that were flowing into the Aral Sea, resulting in the virtual death of that sea. That is now recognised as a great environmental disaster, perhaps the greatest ever, and desperate attempts are being made to reverse it. With the 'linking of rivers' project we may be headed for other unforeseen disasters and may discover this too late. A degree of caution seems warranted before the Government embarks on this enterprise.

Those who advocate caution are apt to be accused of timidity and exhorted to look at China. The fact that the Government of China has been able to push the Three Gorges Project through does not prove that it is a good project. Only the future can tell us whether it is a boon or a disaster. The opposition to Three Gorges in China is muted because dissent is not easy in that country. Those who are envious of China's ability to 'get things done' must reflect on how far they are prepared to go in emulating that system.

Generating New Conflicts?

We have not so far been able to persuade States *within* a basin to share river waters (*e.g.*, the Cauvery Dispute); instead of resolving such *intra-basin* disputes through the better, more economical and more cooperative management of the resources of the basin, should we try to bring water from another and more distant basin? Further, even if we assume that the conflict *within* a 'water-short' river-basin will be eased by the importation of external water, such an effort may initiate new conflicts *between* basins. Several State Governments are opposed to, or lukewarm about, the river-linking project. Bihar,

West Bengal, Punjab, Maharashtra and Kerala are examples. Kerala is stoutly opposed to the proposed Pamba-Achankovil-Vaippar link. There is also some opposition on the part of Karnataka to the idea of the eastward diversion of west-flowing rivers. It may be argued that we should not allow ourselves to be deterred by such political difficulties, but is it really necessary to generate several new inter-State conflicts?

The NWDA proposals include a Brahmaputra-Teesta (or alternatively, a Brahmaputra-Ganga) link, but one wonders whether major diversions from the Brahmaputra are likely. Perhaps we need not discuss this further except to say that the sensitivities of the North-eastern States must be kept in mind. It seems hardly necessary to add one more element of discord in an already difficult situation.

Turning to international aspects, there are serious apprehensions about this project in Nepal and Bangladesh. These may be justified or misconceived, but criticisms of the kind that one heard in earlier years from Nepal in relation to the Kosi and Gandak Projects and from Bangladesh in relation to the Farakka Barrage are being voiced again. Several regional and international conferences are being held at Kathmandu and Dhaka. It is very necessary for the Government of India to explain matters and minimise if not eliminate misunderstandings.

Diversions from the Ganga?

A link between the Himalayan and Peninsular components seems envisaged (Ganga-Damodar-Subarnarekha-Mahanadi). This has caused concern not only in Bangladesh but also within India, in Bihar and West Bengal. When this point was raised sometime ago, it was stated by a senior official of the Ministry of Water Resources that "at no point would waters of the Ganga be transferred to any of the Himalayan or Peninsular rivers." If no transfers are envisaged, there is nothing more to be said. However, the proposals of the NWDA did include some transfers from the Himalayan rivers westwards and southwards. It is that kind of expectation that gives the project its popular appeal, particularly in the south. Here again there is some (perhaps inadvertent) ambiguity. Does the river-linking project include any diversion from the Ganga, and if so, how are the apprehensions of Bangladesh, Bihar and West Bengal going to be set at rest? If no such diversions are intended – at any rate the Himalayan links seem to have been deferred for the time being - how is the Government of India going to disabuse the southern States of the wrong impression that waters from the north will flow to the south in large quantities under this project?

Need for Examination of Individual Links

If we proceed from the 'concept' or umbrella project to the individual links, it is clear enough that in each case there will have to be an examination of the calculation of surplus and deficit, the need for the transfer/link, whether it represents the best out of the available options, its techno-economic soundness and viability, the projected benefits, the environmental/ecological implications and remedial measures, the human aspects (displacement, resettlement and rehabilitation, and the rights and risks of the affected people), the international dimensions (if any), and in the light of all these, a comprehensive techno-socio-economic-ecological analysis of the balance of costs and benefits (including the non-quantifiable aspects) and an appraisal of the project for an investment decision. We do not know what the outcome of that process will be: all projects may pass the test; all may fail; or some may survive a stringent scrutiny while others may not. In advance of that process, a project was announced in 2002 and expectations were raised in the general public. There was reason to fear that this might reduce the whole process of examination, evaluation and clearance to a mere formality, a mockery. That fear cannot be dismissed as unreal, so long as the Hon'ble President of India continues to commend the Project in his speeches, and the Supreme Court continues to monitor the Project.

Pre-empting of Resources

If this mammoth project, not reckoned earlier in the planning processes, is now to be inserted into the Plan, serious distortions may result. There may be a pre-empting of resources, and a distraction of attention from the things that need to be done. Plan outlays are barely adequate even for the completion of projects already undertaken. From the Sixth Plan onwards the stress has been on consolidation rather than on new starts. Against that background, it seems strange to embark on a major river-linking undertaking. The rough figure mentioned in the Supreme Court in this context was Rs. 560,000 crore. That figure may go up in the course of actual implementation, but even if we ignore that point, the pre-empting of resources of this magnitude for this project will render the whole planning process meaningless. We may be wasting a good deal of time in pursuing this chimera, and distracting ourselves from finding time and money for more modest, worthwhile and urgent activities, such as extensive water-harvesting all over the country (wherever feasible) and the onerous but important task of rehabilitation of tanks in the South and other similar traditional systems elsewhere. Even more important is effective demand management through improved efficiency and economy in water use, whether in agriculture or in industry or in domestic and municipal uses, so as to minimise the need for supply-side solutions. These ought to be our priorities, but a preoccupation with the gigantic river-linking project may prevent adequate attention being given to them.

Private Investment? Pricing of Water?

Apart from the pre-empting of resources, the huge costs involved in the linking of rivers and long-distance water transfers will make the water at the receiving-end very costly indeed. There is hardly any possibility of recovering even a fraction of those costs from the users, who will doubtless argue that this is infrastructure development and that the state must bear the cost. However, the possibility of private sector investment is also being explored, and the question arises whether the investors will be able (or should be allowed) to charge full commercial prices. The Enron case comes to mind. Moreover, the question of private sector investment also raises the issue of entrustment of control over natural resources into private (and perhaps even foreign or multinational) corporate hands. These aspects and dangers can only be flagged at this stage. There is not enough information at the moment for a proper discussion of these matters.

Conclusion

It is clear that there are many questions that need to be gone into, and that the necessary information is not in the public domain, or has not yet been generated. Pre-feasibility and Feasibility Reports, said to have been prepared, are kept under wraps. Only one Feasibility Study has been made available so far, namely, the FR of the Ken-Betwa Link. It has been examined from different angles by two eminent academics. Both have found it seriously wanting. The possibility that some of the other links might similarly be found wanting cannot be ruled out.

3

Linking of Rivers: Learning from Others' Mistakes to Avoid Damage to People and Ecosystems[1]

BIKSHAM GUJJA
HAJARA SHAIK

Introduction

Several institutions want to offer different solutions to tackle India's water crisis. Now, the latest "mother of all solutions" that is being presented is inter-linking of the rivers. This is not yet a project, but a concept with its origins in the 1950s. Though there is no single 'project' document called Interlinking of Rivers, the Government has made all efforts to package several old water infrastructure projects where some pre-feasibility work was done to transfer water from one basin to another, as a single 'project' called Interlinking of Rivers (ILR). The earlier chapters of this book have described ILR adequately. The Ken-Betwa link is also based on a decade-old document. This article examines the experiences of such grand water infrastructure projects elsewhere in the world that could offer some lessons for India. The objective of the article is to highlight two main points; firstly, the fact that many grand water infrastructure projects have ignored social and ecological dimensions which need to be taken into consideration right from the planning stage, and secondly that India should take into consideration the enormous costs of these dimensions while assessing the ILR, the costs of which have often not being taken into consideration in the case of other large infrastructure projects in other parts of the world. Therefore, India should learn from others' mistakes well before embarking on this massive project. It is also essential to make it clear that not all the

1. Views expressed in this article are the authors' and not necessarily endorsed by the institutions with which they are associated.

major infrastructure projects are bad. It is also a separate debate altogether whether India needs large projects or not. This article simply presents some of the major water infrastructure projects, not in details of course, but some essentials from where India can draw some lessons. By understanding some of the difficulties of these major projects, perhaps India could save money and lives and could design water infrastructure projects differently to benefit people.

China's South to North Transfer

"The project (South-North diversion) is aimed at relieving severe water shortages in North China," said the then Prime Minister of China, Zhu Rongji, on November 26, 2002. This was the same week the then India's Prime Minister, Atal Behari Vajpayee, said in Parliament that the "Centre plans to take up the scheme for interlinking of rivers on a 'war footing' as a long-term, permanent solution to the recurring problem of drought."

China with 2,800 billion cubic meters (cu.m.) of freshwater resources has the world's sixth largest water resources, measured in volume. When it comes to per capita water, China has 2,344 cu.m. ranking only 88 in the world. Water scarcity in China is complicated by an uneven population distribution. The so-called water surplus region of the South, the Yangtze (Changing) basin and other southern rivers contribute to more than 80 per cent of river run-off yet feed less than 40 per cent of arable land. The water deficit in the North and North-east — the Huanghe (Yellow), the Huaihe, the Haihe basins and the North-west inland region together have more than half of the geographic area with 45 per cent of the arable land and nearly 36 per cent of the population but possess only 12 per cent of the water resources.

The concept of a water transfer from South to North China originated in the 1950s. Mao Ze-dong during a visit in 1952 to the water works of the Yangtze basin reportedly suggested that the North should borrow some of the extra water it needs from the South. After several modifications, three routes were put forward for serious consideration: The Eastern, Central and Western routes.

The project was officially launched in December 2002. The estimated investment needed for this project is 486 billion Yuan or

$59 billion at the current exchange rate. This will be twice the cost of China's Three Gorges Dam, the world's biggest water infrastructure project to date. This new project will divert 45 billion cu.m. of water from the Yangtze basin in the South to the Yellow, Huaihe and Haihe in the North through the Eastern, Central and Western routes.

The total volume of the water transferred by these three routes will be more than 60 billion cu.m from the Yangtze basin. This is more than the natural flow of the Narmada in India. This will be the biggest water transfer project in the world. The average capital cost for diverting each cu.m. is estimated at about $1.2.

China's South-North scheme has met with much criticism within the country. While it is true that China is facing water shortages, these can also be met by proper water management. Studies have indicated that with smaller outlays invested in waste reduction and wastewater collection and treatment programmes, China could save up to 50-80 billion cu.m. of water more than the proposed water transfer.

A detailed financial analysis of the real cost of the water in the Eastern route has revealed that the water will only be affordable by urban and industrial users. Similarly the cost of the water of the middle route will only be affordable by industries. Such massive water transfers may further lead to wastage of water and generation of pollution The much needed investments in water savings and pollution control may not be made because of this massive project.

Aral: Symbol of Planned Destruction

The drastic changes in the Aral Sea — area and its quality, is perhaps one of the best known examples of the large-scale water transfers conceived in the 1950s and which featured in the Soviet-era plans. The Aral Sea was the fourth largest freshwater lake in the world with an area of 66,000 sq. km including 2,200 sq.km of islands. The lake's water volume exceeds 1,000 cu. km (as much as the total annual discharge of Ganges (493) and Brahmaputra (510) together) and its average depth is 30 meters. In the 1960s, the Aral Sea had a thriving commercial fishery and was a popular recreation spot. Just in one year from 1995 to 1996, the lake area decreased from 3885 sq.km and volume by 36 cu.km. Salinity of water increased to 46 g/l,[2] much higher than the sea water.

Two rivers, the Amu Darya and the Syr Darya flow into the Aral Sea, the former is the larger with a basin area of 309,000 sq.km (size of the Godavari basin) and total length of 2500 km. The mean flow of both the rivers is about 116 cu.km (almost same as Godavari).

Until the 1960s, the former Soviet Union used mostly primitive irrigation techniques. As part of the "integrated method of development of desert lands," this included moving water and people to the fertile areas.

It was a conscious decision to divert almost the entire flow of the rivers to other uses. And this was true of the Aral basin as well. About 92 per cent of the diverted water was used for irrigation and the rest for industrial and rural water supply. The irrigated area in the Aral region expanded from five million ha to 7.9 million ha.

The consequences of this water diversion are highly visible. The Aral lake area has been reduced to almost half (38,000 sq.km) and the volume of the water in the lake has fallen by one third (354 cu.km). This exposed the lakebed, which became dry and salinity levels in the lake increased, almost eliminating the commercial fishery which at its peak, produced 40,000 tonnes per year and supported 90,000 local fishermen.

Downstream provinces were affected by highly saline water and wind blown particles. Almost 3.5 million people living in the area were affected in one way or other. This region is often called a "disaster zone."

One report by the World Bank said of the Aral Sea "the large-scale diversions created an ecological and human disaster. Increasingly saline soils reduced agricultural productivity resulting in some of the worst poverty in the region."

Studies estimate that economic damage due to such large-scale water diversions is between $1.25 to $2.5 billion a year and this is just to human health, tourism and agriculture. More detailed studies

---

2. Information on Aral sea its changes since 1950s, data on water use and land use pattern can be seen available *http://www.dfd.dlr.de/app/land/aralsee/*

and the estimation of long-term implications could increase the cost even more. The consumers of water paid nothing for what they got and today several projects are being launched to control the damage through several national and international initiatives. But it is most unlikely that such attempts at damage control will ever restore the Aral Sea to its healthy state and productive use of the pre-I960s.

The now rusty boats in the Aral Sea serve as testimony to the ecological disaster of water management based on the 1950s understanding, "never let a drop go waste to sea."

Recently many projects have been initiated to improve the situation. Hundreds of millions of dollars are being spent to address the issues. The Aral Sea Programme supported by World Bank with close to hundred million dollars is still going on. The World Bank reports that "Although improvements are already noticeable, environmental degradation cannot be reversed overnight, and the project team estimates that sustainable results will only be obvious about 10 years after the project is completed"[3]. The recent increase in oil prices improved the Kazakhstan economy. The country has initiated a massive ecological restoration project to repair part of the damage. The objective of the project is to double the divided Aral Sea in Kazakhstan. This is going to cost a lot of money and efforts. The project builds on the earlier initiatives of the World Bank. The focus is on "What man has destroyed, man can now restore."[4] There are many technical and practical difficulties, but for now many fishermen are hoping to come back to fish in the Aral Sea. Although, the restoration will never be complete, any improvement in the situation is welcome. However, for now the Aral Sea is still sinking and the cost of partial repair is going to be high, and might be more than the so-called benefits of the Aral Sea water diversion. And so, "For most people, Aral today is a symbol of how badly the Soviets treated Kazakhstan" (IHT, Sept. 2005).

---

3. 'Saving a Corner of the Aral Sea' report in World Bank *website http://www.worldbank.org.kz/WBSITE/EXTERNAL/COUNTRIES/ECAEXT/KAZAKHSTANEXTN/0,,contentMDK:20633763~menuPK:361875~pagePK:1497618~piPK:217854~theSitePK:361869,00.html*

4. "Kazakhstan plans to restore a blighted sea" by Christopher Pala International Herald Tribune, Wednesday, September 28, 2005.

Irtysh-Karaganda (Satpaev's) Canal

The Irtysh is the principal source of water for 16 million people in Kazakhstan. The country is searching for new opportunities for progress and new alignment to achieve that. In order to achieve these goals, this country has to deal with the past. That past is linked with the Irtysh-Karaganda canal also known as Satpaev's canal. This huge water transfer project has now collapsed.

As part of the massive agricultural expansion during the Soviet era, the cultivated land increased from 7.8 million ha in 1950 to 28.5 million ha by 1960. In 1993, 90 per cent of the cultivated land was still under state farms *(Sovkhoz)* and collective farms *(kolkhoz)*.

The Irtysh and its tributaries are the main source of water for North-Eastern Kazakhstan. The river is one of the longest in Asia and travels 4,284 km. The river originates in Xinjian (inner Mongolia) of China and travels 580 km. in the country before entering Kazakhstan. After flowing 170 km. in Kazakhstan the river flows into the Zaisan Lake. This part of the river, until Zaisan Lake, is called Black Irtysh (Chernyi Irtysh). Then it becomes the Irtysh and the Northwest.

The Irtysh-Karaganda canal was part of the scheme to expand irrigated land in 1960s. The canal was constructed between 1962 and 1974 to divert the Irtysh water to the Nura of the Karaganda region. It is a massive water diversion scheme in Soviet style. The underlying rationale is the same as the others: water is free and it can be transported anywhere and if you do not do that the water is wasted. The Irtysh is more than 450 km. from Karaganda and its level 450 meters lower, calling for pumping of water. Canals 458 km. in length, 22 pumping stations, 14 dams, 17 bridges, approach roads of 524 km. and transmission lines of 703 km., were built for the project. Total annual energy consumption was 2200 million kWh. The total cost in the 1960s then was 338 million roubles (about $1.5 billion). What are the intended benefits of the canal? Water for irrigation, municipal use and industry. About 400,000 ha of land was supposed to be irrigated.

In 1986, the canal provided 772 million cu.m of water, about half of it for irrigation. Officially it was reported that about 278 million cu.m. of water was lost in transportation, which included 220 million cu.m of evaporation during the summer.

Almost one-third of water was lost in transit. The intended capacity of the canal was 2,400 million cu.m, which was never achieved. Water was used to grow potatoes, vegetables and fodder. Even in Soviet times, farmers paid nothing for water and there was no incentive for conservation. Later, due to contamination of the Nura with mercury, water use was reduced.

Currently there are three types of issues with this canal project — social, economic and ecological.

Due to political changes in the region, the massive KazKarmet steel works at Karanganda has collapsed. The steel plant supported the operation of the canal. Now there is no money for maintenance. The industry, which was using the water from the canal, vanished. Changes in the farming policy and land privatisation also resulted in major changes in agriculture in the region. Farmers are no longer able to pump the water from the main canal. They cannot afford it so water use for agriculture too has reduced considerably.

Water pricing calculations indicated that the variable cost is $0.26 per cu.m and with fixed costs the cost may go as high as $1 cu.m. A recent World Bank report states the unsustainable nature of irrigation in the Kyzyl Orda province in the south Syr Darya Basin; "the actual price of water was fixed at $ 0.06/m although the real cost of *O&M* (operation and maintenance) would require $0.56/m. This low price has not stimulated farmers to adopt water saving techniques."

Ecological problems are far too many. They were not even on the agenda when the canal was planned. Weeds, canal bank instability, salination, degradation of soil, invasion by alien species are some of the ecological problems. At present, the Irtysh-Karaganda canal is not operating. Only occasionally are the pumps operated so that they will not rust and become totally useless. There is a move to restore the system.

More recently, issues related to sharing of water with China have come up. China too has claims on Irtysh water. China wants to divert 1000 million cubic meters of water in the next 20 years. The Chinese negotiations for sharing of water are also linked with its $9 billion oil pipe line to bring oil from Kazakhstan to Xinjiang province.

But it is unlikely that the Kazakh Government will find the money required. Even if they did do so, operating costs are too prohibitive for farmers to pay for water. One more water engineering achievement of that time has collapsed on its own.

GAP-Turkey: Bridging the Gap?

The South-eastern Anatolia Project, known as GAP in Turkey, is another major water infrastructure initiative currently being implemented.

The Euphrates and the Tigris originate in the North-eastern Anatolia region of Turkey and flow into Syria and Iraq before flowing into the Shatt ai-Arab marshes and to the Persian Gulf. Turkey contributes to 89 per cent of the Euphrates flow and 52 per cent of the Tigris, and wants to use 52 per cent and 14.1 per cent of the waters respectively. Being the upstream country and also the major contributor to the waters to both the rivers, Turkey has the advantage of being able to use the water and also feels that it has a legitimate claim on the waters of the two rivers.

As part of GAP, 12 water infrastructure projects have been prepared — six in the Euphrates basin and six in the Tigris basin. This includes building 22 major dams, and 19 hydropower plants. After completion, the infrastructure will provide irrigation to 1.7 million ha of land and have 7,476 MW of installed hydroelectric power capacity with 27 billion kWh of annual energy production. It was also estimated that GAP will contribute to a five-fold increase of the region's income. The entire project is estimated to cost $32 billion at the 1997 exchange rate. When they are completed 29 per cent of the rivers' water will be managed through this project and 20 per cent of the country's irrigation will be in the project area and 22 per cent of energy will be generated by GAP. The project was scheduled to be completed by 2005 but now has been extended to 2010.

By 2001, more than 48 per cent of the money was spent and another $2 billion a year is required until 2010 to complete the activities. As against the estimated 1.7 million ha of irrigation expansion only 215,000 ha, just 12 per cent of the total, has been completed.

While Turkey is determined to complete this massive project by 2010, it is spending around seven per cent of the country's investment allocations on one single project since 1990. Though the project has crossed half way in terms of spending and two-thirds of the scheduled time, it is nowhere near completion. There are many problems with the implementation as well as the projected benefits.

The problem of salinity has already started. It was mentioned in the annual report of the GAP that more than 29,000 ha of land is affected by salination in the Sanhut-Harran plains. Of every three ha of new irrigation created in this region, one ha began to show salinity problems. The cost of irrigation is also very high. Not even a fraction of the *O&M* costs are being recovered from the farmers. The farmers, who anticipated water, are totally disappointed and selling their land. Local people who are supposed to be benefited by the project are complaining about new infectious diseases.

The GAP project also has health dimensions. Experts predict that this project might lead to many infectious diseases such as Leishmanaiasis, amebiasis etc. and that the 'cost of combating diseases can be very large.'[5]

The project is still being constructed, and it is already running several years behind schedule. Turkey is spending lot of its own money and raising more from various sources to complete the project. Already the project has proved expensive and the projected benefits are far exaggerated.

Large Infrastructure Projects: An Outdated Concept?

These large-scale water infrastructure projects were part of the socio-political situation prevailing in different parts of the world. Things have changed radically in many countries. The concept of development is much different now. These projects offer major lessons. Learning these would save money, lives and ecosystems without compromising on developmental goals.

Not a drop should go waste into the sea: All these projects are based on concepts which evolved during the 1950s. At that time, the U.N. promoted such projects as part of 'stability and peace.' That was the time when big projects and technology were seen as the answer to poverty. It was also the time when many countries, after gaining

5. Group of health experts published a letter in 'Emerging Infectious Diseases' in April-June-1995.

independence from colonial powers, wanted to express their national confidence through such major projects. Water was not yet a problem, but bringing it to people was the issue. It is also wrong to say that these projects did not benefit people. Indeed they did provide benefits. But the costs were too high and, as time passed, the costs exceeded the benefits. Also it was the poor who paid the costs and others got the benefits if not always, then most of the time. Though, these concepts are out-dated, many politicians and hydrological engineers still believe that water let into the sea is being wasted — hence the engineers' slogan "Not a drop of water should go waste into the sea."

Water is free and cost recovery is not even on the agenda: When these schemes were designed, financial recovery was not even on the agenda. Soon some projects became a major burden on the state. Often the operation and maintenance costs far exceeded the benefits. Here again the water users were not concerned, since they were largely getting water free.

Environmental aspects ignored: Environmental costs were not considered then. Many countries did not have any government structures to deal with the ecological problems raised by these projects such as salination, water pollution, sedimentation and water logging, which have now become major issues. Often it was too late to do anything and the costs of addressing these issues were considered a luxury for poor countries.

Poor can be shunted around: Centralised, top down planning was then the method of governance. People were to be mobilised, shifted and displaced in the greater cause of the country. People had very few rights and political power.

It is also clear that the current water crisis is the result of the earlier bad investments and management in many countries. More of the same might further aggravate the crisis. Institutions like the World Bank, which actively promoted such large projects as solutions are now cautious about them.

Lessons for India

The interlinking of Indian rivers proposal originated at the same time in 1950s, when the world leaders were fascinated with large water infrastructure projects. To some extent those large water infrastructure projects were statements by several countries about their technological advancement, social organisations and taming the nature for human benefits. That was the understanding and perhaps needs of those times. Those large water infrastructure projects did play a major role in addressing the challenges of those times. Indeed some of those projects did contribute significantly to improve the foodgrain production. But times have changed, the cost of those large projects are being evaluated. Some of those projects such as mentioned above have become great liabilities on the nations. Some of these projects were built under different conditions for example:

a) People, particularly poorest of the poor had little rights then. Millions of people were just asked to vacate their lands, homes in the name of public good. Those people were never properly compensated. Even the nominal costs, and social costs of such displacements of these large projects were never computed and included in evaluating the cost-benefits of the projects.

b) The concept of ecological damage and cost of destruction of ecosystems were not even on the global agenda then. Therefore, while designing and constructing these large water infrastructure projects the ecological dimensions were never taken into consideration.

The costs of these large water infrastructure projects were too great. That was a different time when they were planned and the political, social and economic logic was completely different. The logic of water management has radically changed. Like in any other infrastructure project a new way of designing and a new way of computing costs and benefits have come into force. The Task Force constituted by the earlier government was expected to study ILR with present standards and guidelines. They were supposed to take into account the experience of the other countries. The then Task Force Chairman Suresh Prabhu publicly declared several times that they were open to suggestions. Now the Task Force is no longer in existence, and there is no information about what the Task Force has done; no report or key documents were made public. The current government has certainly made some documents public, but many more are still not available. Certainly the government has not taken into consideration learning from international experiences. At one meeting Mr. Prabhu then Chairman of the Task Force, had said, "If the links cause more damage to ecology than bringing benefits, then they will be abandoned." But the question is, who has conducted such cost-benefit analysis, have they taken into consideration the experiences from other countries? The answer so far is no, no such analysis have been done. Or at least there is not a single critical document in the public domain, to explain the cost-benefit analysis, the ecological damage likely to be caused by the project and the strategy planned to mitigate such damage. India can learn from international experience, which shows that similar projects turned out to be far more expensive than originally estimated; they were never completed in full; there were unexpected implications for human health, soil, sedimentation and resultant pollution could be much more than the benefits; they have taken much longer than estimated; these projects have often impoverished local people and lastly the alternatives were a lot cheaper and socially more acceptable.

Conclusion

ILR proposed by the government as an answer to India's water problems is based on concepts in the 1950s. India is not the first country to contemplate such huge projects. There are many examples the world over about how such major water infrastructure projects have proved expensive and led to massive damage to ecosystems and economy. There are many studies about such cases. India, by examining such cases could draw some important lessons. After examining all the dimensions of ILR, or parts of it, the government can prepare detailed project reports (DPRs). Such DPRs can include ecological, social dimensions of the proposed linking projects. Those documents can be released to the public for further discussion so new ideas and new concepts can be incorporated. As part of preparing DPRs, India can constitute a technical panel to study some of the international projects which are in the pipeline and which are completed to draw some lessons. More information, more analysis on such large projects can only benefit the country. The ILR is huge, therefore, it should take into consideration all aspects of this project including learning from others, to avoid all possible mistakes in the interest of the people and ecosystem. The Government has an obligation to prove to the nation that they are not doing something, the consequences of which the future generations will have to face. Many of us may not be around to see the completion of this "project", even if it has been started, but our children may have to pay with their future such as in the case of the Aral Sea and Karaganda Canal.

References

Aksoy, Serap; Ariturk, Sedat *et al.*, (1995) "The GAP Project in Southeastern Turkey-Potential for Emergence of Diseases" in *Emerging Infectious Diseases* Vol. 1 no. 2 April-June 1995.((ftp://ftp.cdc.gov/pub/EID/vol1no2/adobe/letters.vol1no2.pdf).

Asia Times (1999) "China Discusses Future of Irtysh River" by Bruce Pannier and Edige Magnuin.

GEF (1998) *Aral Sea Basin Programme* (ASBP)-Water and Environment Management Project

International Herald Tribune, 28 September 2005 Kazakhstan plans to restore a blighted sea, by Christopher Pala.

Olcay Unver (2002) *Turkish Experience in Regional Development Based on Water Resources: GAP* Regional Development Administration Ankara, Turkey.

River net (2002) *NGOs Welcome European Parliament's Clear "No" To Water Transfers. http://www.rivernet.org/prs02_02.htm*

The River Naura in Kazakhstan: http://ivl.dataphone.se/twinbas/nura.asp

Tony Long (2003) *Spanish Hydrological Plan (SNHP) and EU Law and Policies- Submission to European Paliment* on 6th May 2003.

UNDP (2004) *Water availability in Kazakhstan. http://www.undp.kz/library_of_publications/files/2496-13191.pdf*

Warren S.C. (2001) "The Proposed South-North Water Transfer Scheme in China – Need, Justification and Cost." *WWF Yangtze Programme Report 15*, WWF China Programme Office, Beijing, China.

WWF (2003) *7 Reasons to Oppose SNHP Suggested Actions and WWF's Alternatives.*

What is Gap, General information on GAP and Ataturk Dam: http://www.adiyamanli.org/gap/info_on_gap.htm#WHAT%20IS%20GAP

World Bank (2001) *Syr Darya Control & Northern Aral Sea Phase I Project*. This project is continuing and expected to close in 2007. Project information document (PID) for May 2001 is available on the world Bank website.

Yu Xiubo (2001) "The South-to-North Water Transfer Project in China: Issues Should Not be Ignored'- *Report of Chinese Academy of Science Innovation Projects* (No. KZCX2□|405). Institute of Geographic Sciences and Natural Resources Research, Chinese Academy, Beijing 100101, P. R. China.

This is a revised version (with references) of the article published by the Hindu in 2003.

Section 2

Ken-Betwa Link

4

Methodology of Irrigation Planning

The Ken-Betwa Case

YOGINDER K. ALAGH

Introduction

The Ken-Betwa is a case study of the Interlinking of Rivers Project planning methodology as a published transparent document. This is a part of the effort of the authorities to encourage debate on the project and this transparency is to be commended. It is important to note that this is the first time a pre-feasibility project document and the methodology it is based on has been made available for widespread discussion and this is a breakthrough. The only case when a document of a similar kind was made available was the Sardar Sarovar Plan but that document although in greater detail was made available after the plan was approved. This paper takes the attempt of the authorities seriously and discusses the document. It addresses the following issue:

- *Does this study meet the criteria of adequate planning methods for river basin transfer projects? The test can be state of art project design in India or the ILR methodology requirements itself. A stricter test would be global standards, but we do not use that criterion.*

We examine in this paper the following aspects:

- Resource inventorying; agro-economic zoning,
- Water Demands; particularly for agriculture in each zone,
- Conjunctive Use; alternative methods for delivery design,
- Demand driven project design methodologies for water delivery systems, and
- If adequate we look into hydrology, storage design and rehabilitation policies; if not go back to project design.

Resource Inventorying and Agro-Economic Zoning

There is a synergy in water as a resource emerging from complementary and sustainable use with more holistic approaches and this can only be captured with more integrated methods of measurement. An example is the simultaneous use of tanks and surface conveyance strategies. More precise assessments can in many cases make the difference between scarcity and conflict, on the one hand, and the possibilities of a judicious management strategy on the other.

Water is both surface and ground. These concepts have been described well by geographers and a classic description is by the Russian academician Dr. Galina Sadasyuk (See Alagh, 1991 for a detailed discussion). G. Sadasyuk and P. Sengupta, (1968) for example, divided India into 18 agro-climatic zones and 44 sub-regions. (See Alagh, 1988 and Planning Commission, 1989). Similar exercises have been developed in other large countries, for example, Brazil, Indonesia and even France (See Ignacy Sachs, 1991, Lutfi Nasution, 1993).

It is at the agro-climatic sub-zonal level that resource availability of water can be given teeth to. This is not to say that the existing estimates of a more global kind are of no use. In many cases they are all we have, but the quest for improvement is not only that of conceptual and numerical precision, but that of an improved understanding of the processes through which renewable supplies of water are generated and without that, numbers of x cumet/per caput will remain not only an abstraction, but also subject to a degree of avoidable variability. A parallel in Economics was that the measurement work of Wassily Leontief and Richard Stone was closely intertwined with the conceptual development of modern macroeconomics.

There has been some debate around these concepts. Discussing water, K.V. Raju and Tushar Shah recognise that "The physical diversity in India is further accentuated with a definite variation in soils and cultural mores. Agro-Climatic Regional Planning (ACRP) presents itself as an alternate approach for classifying water resources in the country. This method of categorising and classifying water resources employs a utilitarian perspective by subsuming the entire range of "land and water resource based activities under the broader range of farming systems" (Planning Commission, 1996)." Following earlier studies, described earlier, (Alagh, *et al.*, Planning Commission, 1989) they state "The ACRP strategy aims at a more scientific utilisation of available resources by taking a holistic view of climate, soil type, topography, water resources and irrigation facilities and relating them to output and employment (Planning Commission, 1989). As a first step in the ACRP approach, the country was regionalised into 15 agro-climatic zones having 73 sub-regions. The zones are delineated so that the regions within the zone have a high degree of commonality on a number of factors ranging from climate, rainfall, water demand and supply characteristics, aquifer conditions to soil types and topography."

However to them "The ACRP approach has its perceived shortcomings, which may be due to the focus of the approach on agricultural resource planning. According to this approach, technology and sustainability (which are projected as the two dimensions integral to resource based planning) are defined only within the constraints placed by finance, economic efficiency, institutional and social imperatives." This is on account of "Environmental and ecological constraints placed on water resources to maintain ecosystems services and also to service the needs of non-human life forms are conspicuously absent from this approach. Furthermore, the basic unit of planning within the sub-zone according to the ACRP approach is the district, which is also a political and administrative unit. Sub-zones and zones bring together different planning units which often cross state boundaries. The prospect of coordination on water resource issues (that seldom conform to political boundaries) between different departments, both within the state and in some cases across different states, raises doubts about the successful implementation of this approach (Nadkarni, 2001). Nadkarni suggests the evolution of an agro-ecosystem policy as a more relevant local planning tool for achieving the twin goals of sustainable use of natural resources and for ecological sustainability (Nadkarni, 2001). Agro-ecosystem policies, as and when evolved, should address the issue of water resources in a holistic framework."

Box 4.1

Dublin Statement

- Water is a finite, vulnerable resource essential to sustain life, development and the environment. The effective management of this source demands a holistic approach linking social and economic development with protection of natural ecosystems.
- Water development and management should be based on a participatory approach with decisions taken at the lowest appropriate level.
- Women play a central part in the management, provision and safeguarding of the resource.
- Water has an economic value in all its competing uses and should be recognised as an economic good.

Source: UNDP 1998.

To Raju and Shah, and like minded scholars, "River basins, as is evident from the definition, is a natural division that has physical existence (unlike political and administrative boundaries). Furthermore, river basins occupy the apex of a hierarchy of natural sub-divisions like sub-basins, watersheds and micro-watersheds. This is perhaps, the most powerful rationale for the introduction of management of water resources through a river basin approach."

The present author when Minister of Planning had set up The National Commission for Integrated Water Resources Development in 1996, a Blue Ribbon Commission as he called it. Unfortunately the Report of this Commission remained a classified document for long and was not known, but has been released. Raju and Shaw describe that "the (NCIWRDP) devised a new classification that categorised India into 24 important basins. Of these basins, 12 basins are designated as major basins (i.e. basins with areas in excess of 20,000 sq. km.). Figure 1.1c (Annexure to the Report) indicates the river basin map of India. From the basin map it is evident that the Ganga-Brahmaputra-Meghna sub-basin is the largest river basin in the country with an area of 1.097 Million km^2. The basins of the east coast are significantly bigger than those of the west coast. It is also significant that the geographical location of the Himalayan basins coincides with the location of India's Green Revolution – that required intensive agricultural systems, HYV of crops, extensive canal irrigation networks. Another significant point is that all the rivers of peninsular India are non-perennial and consequently cannot depend on rivers as a perennial source of irrigation water. Therefore, alternate sources of irrigation like tanks are prevalent in peninsular India." The NCIWRDP, however recognise that for different aspects of policy making, States, agro-climatic regions, Districts and micro-watersheds could be a unit.

On a more general plane, it has been pointed out as a part of the global discussion on emerging research issues of the new millennium, emerging from the UN Millennium targets (See the document on emerging issues in social science research published by the Scientific Steering Committee of the International Social Science Programme called MOST-Management of Social Transformations: A. Ruijter, *et al.*, More on MOST,) to the effect that concepts of water stress and water scarcity have been developed based on an index of per capita fresh water needs (Malin Falkenmark, 1991) are of an average ready reckoner kind, with the considerable advantage of near global applicability, but needs more work. It may be seen above that Indian scholars have also used different classifications for work-dry regions, agro-climatic, etc. The issue of the unit of analysis is therefore open and still needs work.

While not strictly so, water can be treated as a renewable resource. Fresh water supply can be determined in an agro-climatic regime in a defined time period. For a political entity like a country or smaller administrative boundaries, a rough water availability indicator can be estimated by dividing the renewable water supply in a time period by the population of the country or region, in the same time period. The available renewable water supply includes (a) internally produced surface water, (b) internally produced ground water, and (c) river flows from the adjoining countries of the region. The Falkenmark Index is based on a yearly time period and consists of 1000 persons per 1 million cumet or one flow unit. 1000 cumet/per caput annual, therefore, becomes the dividing line for water scarcity (Falkenmark, *et al.*, 1989). This norm has been used in the global discussion. Countries in the range 1000-1700 cumet/per caput annual are classified in a water stress category. Below 1000 cumet/per caput annual, the water stress index is that of extreme scarcity.

These indicators are admittedly of a rough kind. Water flows do not follow administrative or political boundaries. Aquifers are confined by geo-physical and not political and administrative features. Climate also determines the demand for water, including evaporative demand. Seasonal rainfall cycles can be very different across countries. For structural understanding and policy analysis, general indices have to be used with care. (See Y.K. Alagh, 2002, in UNESCO, A. Ruijter, *et al.*, op.cit).

Water availability, in fact is now also estimated as a part of structural analysis and advanced project level design for both aquifer management and drainage and conjunctive use projects, with a reasonable degree of accuracy. For example, the Mike She Model of the Australian Murray Darling Basin Plan models is used by the New South Wales Department of Land and Water Conservation. While it is used for integrated catchment modelling, for the basin, it's main features "are that it can represent all major flow features in an area and is well suited to describe the dynamic interaction between the surface and subsurface water systems" (Peter Millington, 1996). The flow chart of a simple multi-level model can be as follows:

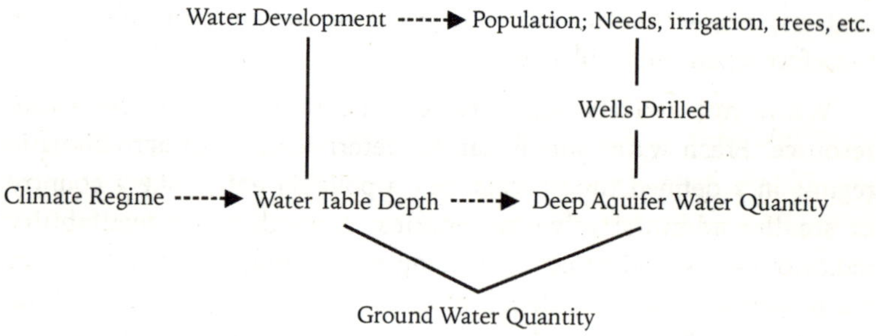

Source: Nicholas Sonntag (1996).

A more detailed version available, which gives detailed additional resources of water made available by a more precise basin level assessment has been reported in a set of studies compiled by William Fisher (1995). For the Mahi Narmada Doab, a model developed by Alagh and Buch, gives initial heads in 1980, calibrated heads for 1990 and forecasts for 2000 under alternative use assumptions. The model shows sensitivity of outcomes to trees planted, surface irrigation and pumps drilled. (Fisher, Columbia Seminar Series, 1995, pp. 291-318).

Interesting work includes sensitivity of estimated resource flows of water available with integration of surface flows with local small storage projects. Alagh (2000, World Water Council Plenary Speech) has reported an augmentation to the extent of 14 per cent of estimated water availability as follows:

Tank Storage in Shedhi System

Year	*No. of Tanks Deepened (Progressive)*	*New Capacity Creation MCM*	*Range of Deepening (m)*
Season 1993/94	150	3.5 (6.0 to 9.9)	1 to 6
By June 1994	254	6.0 (7.9 to 13.9)	0.25 to 9.3

Fourteen per cent additional water cannot be considered trivial in a water short context. It is interesting that the Shedhi Branch of the Mahi system was planned on the basis that there are no tanks in the system.

The point of these micro studies and approaches to assessment of available water resources is to emphasise that the quantities themselves are sensitive to detailed resource inventorying as compared to initial assessments based on empirical formulas, projections of a preliminary kind and in some cases 'guesswork'. This kind of limitation is being addressed, for example by international programmes, etc. But there is a more basic problem. There is a synergy in water as a resource emerging from complimentary and sustainable use with more holistic approaches and this can only be captured with more integrated methods of measurement. An example is the simultaneous use of tanks and surface conveyance strategies. More precise assessments can in many cases make the difference between scarcity and conflict, on the one hand, and the possibilities of a judicious management strategy on the other.

The FAO, for example, moved over from irrigation as an input in the agricultural sector to a perception of the synergistic role of water in the process of agricultural growth as early as the late-seventies of the last century. Thus, the following statement illustrates:

"The pervasiveness of such influences is suggested by a comparison of targets for 1975 proposed in the 1969 FAO Indicative World Plan for Agricultural Development (IWP) with actual experience upto 1976. While actual production growth in most developing regions was below proposed growth rates, most identified inputs.... actually exceeded growth rates proposed for them "(FAO, AT 2000). For example in the period, the targeted growth rate for agriculture in the Asia and Far East Region was 3.7 per cent annual, but the achieved rate was 2.6 per cent. However, arable land grew at 0.6 per cent annual, twice the target rate and irrigated land at 2.2 per cent while the target was 1.8 per cent. Since then the policy and planning exercises have been in terms of more holistic approaches of simultaneous development of land and water. Thus the planning methodologies of AT 2000 considered the development sequence of moving land from a soil and water regime to an alternate one. There were in the numerical models five land and water classes. Moving from bad soils with low and variable water availability to good irrigated lands. This was a more synergistic view (developed from FAO's vast field experiences) of water development preparing agro climatic regimes to adopt specific (appropriate?) cropping and farming systems. AT 2010 and now AT 2025 carried this process to a logical conclusion of sustainable development of land and water resources providing the basis of farming systems. (See Alexandratos, 1995 and Bruinsma, 2005).

The argument here is that conceptually, the measurement of water availability has to be in the context of a physical system and therefore is in the ecological domain.

Regionalisation for Irrigation Planning

Let us begin at the level of the agricultural field. Irrigation engineers have to learn to design water delivery systems differentially for the different regions of a canal-command. For too long, commands of large irrigation projects have been treated as homogenous and uniform entities. Soil conditions, temperature and its distribution, rainfall and its distribution, the ground water regime, existing forest cover and existing tanks and minor rivers and drains are all features which need to be paid very detailed attention. This will normally require that a command is regionalised into components. In a large project in Western India where while the initial intention was to provide for a uniform pattern of irrigation through the command, after detailed consultations with geologists, demographers and land use planners, meteorologists and agricultural scientists, the command was broken up into 13 regions (Sardar Sarovar Narmada Command; See Y.K. Alagh, 1991, pp. 160-162). This phase is important for irrigation planning. This phase has to be seen not just as a question of regional studies. The association of the irrigation engineer with the wider disciplines of geography and land use, meteorology and ground water will from the beginning sensitise the irrigation planner to the diverse agro-climatic and agro-ecological environment in which the water system has to operate and this sensitivity is extremely important to planning the physical features of the irrigation structures.

Ken-Betwa

The Ken-Betwa study does use the word agro-climatic once (p. 77). The methodology paper for ILR project preparation (TOR) demands its use, (pp.22-23, Sec.5.6) but it is not there in the Ken Betwa case. In the TOR also agro-ecological and socio-economic data is required to assess the impacts of the project: not to design it – a typical engineering mindset quite faulty and out-of-date as we saw above. To repeat without an understanding of the soil rainfall regime in detail, canal commands and delivery systems cannot be designed of any sensible kind. This is a serious limitation of ILR planning, which by itself makes it of little utility.

The agro-climatic characteristics of the region are well known as it is a part of the Zone 8 called the Central Plateaux and Hills. This area of 37 mn. Hec. is described as one of the most difficult to plan for in the Planning Commission's agro-climatic plan. (Details in Planning Commission, 1989 and not repeated here) NSA (Net Sown Area) is 17 mn. hec. and 12.8 mn. hec., need soil and water conservation by integrated watershed development. 9.53 mn. hec., have acute soil problems which need treatment.

These problems have been hinted in the Feasibility Document, but are tucked away. The degree of casualness with which resource and human problems are treated in purely engineering kind of planning are shown in the following statements:

- "Occupation

  Agriculture: about 15 to 20% (?) of the people to be affected in the submergence are having agriculture as their occupation. Also, most of the people to be benefited in the command area are dependent on agriculture, either cultivating their own lands or working as labourers in agriculture fields.

  Other than agriculture: The majority of the population in the submergence area is dependent on forest products as the reservoir area is covered mostly by forests. A few of them(?) are either agricultural or industrial workers" (Feasibility Report, henceforth FR, p.5; emphasis and interrogation marks added.)

The Ken-Betwa study has soil information not used for irrigation planning.

Soils

	Category	Per cent area in the sub-basin
i.	Soil on hill and hill ridges (Entisols)	7.3
ii.	Plateau soils (Entisols, Inceptisols and Alfisols)	44.0
iii.	Pediment soils (Entisols and Alfisols)	2.1
iv.	Soils of level alluvial plain and undulating flood plain (Inceptisols and Vertisols)	43.6
v.	Soils of Dissected flood plain Inceptisols)	3.0

A brief description of these soils is given in the following paragraphs.

(i) Soils on hill and hill ridges (Entisols): These soils are fine loamy to coarse loamy in texture and greyish brown to dark reddish brown in colour. *These soils are highly erodible, excessively drained, stony and gravelly in nature. These soils are of shallow depth with low nutrient status and are slightly acidic to neutral in reaction. They are low in organic matter content and have poor water retention capacity. These soils are unsuitable for normal crop husbandry. However, they may be utilised for the purposes of forestry, pasture development and growing grasses. (emphasis added FR, p.31)*

(ii) Plateau soils (Entisols, Inceptisols and Alfisols): These soils occur on level to gently undulating terrain. They are yellowish brown to dark brown in colour. *These soils are shallow to deep coarse with medium textured and low nutrient status. These soils are moderately eroded. Their water retention capacity varies with the soil texture and organic matter content. In this category too, the shallow soils are unfit for normal crop husbandry whereas deep soils are suitable for growing kharif crops like sorghum, millets, pulses, and bajra. (p.31).*

Contd...

...Contd...

(iii) Pediment soils (Entisols and Alfisols): These soils occur on gently to undulating gneissic terrain comprising the vast pediment and piedmont plains characterised by intermittent rocky wastelands and shallow water bodies. They are shallow to deep, coarse to medium in texture and poor in nutrient status. *They are moderately eroded. The deep soils in this category are fit for cultivation of crops like sorghum, til, bajra, millets and groundnut. Shallow soils are best suited for raising pastures. (p.31)*

(iv) Soils of level alluvial plain and undulating flood plain (Inceptisols and Vertisols): These soils occur on level to undulating terrain. These are deep to very deep, fine to fine loamy, well to moderately well drained and contain calcium carbonate deposits. They are neutral to slightly alkaline in reaction. They are low in organic matter content, fairly rich in nutrient status and are moderately eroded. *These soils are best suited to crops like jowar, arhar, soyabean and moong under dry land and sugarcane under irrigated conditions. Wheat, gram and linseed may be grown during the rabi season. (FR p.32).*

(v) Soils of dissected flood plain (Inceptisols): These soils are very deep, well drained, calcareous, and gravelly with abundant lime nodules in the soils. These are fine loamy in texture, yellowish brown in colour with low organic matter content. They possess moderate water retention capacity. *These soils are severely eroded during floods and need extensive conservation measures. Lands situated away from the streams can be put under cultivation for kharif crops viz. jowar, bajara, and arhar. (FR,p.32).*

In the operative details, the geologist is almost entirely bypassed in the project. All these terrible warning signals are ignored in the planning design. This kind of "planning" without consideration of resource costs is unthinkable in 2005.

Planning Delivery Systems: Farmers driving Engineers

The systems we have suggested which are now in operation are first to work out water requirements including drainage and ground water in a level of regional agro-climatic space and then design the irrigation delivery structures. The engineering aspect impacts on economy and society and, therefore, there is the requirement of a wider range of skills to be brought to bear on the solution of purely technical and structural problems. It is true that we have in the past not been the best conveyors of water. Our skills in dam designing and construction are recognised globally, but questions of planning and management of water conveyance system have received much less attention. Conveyance losses are high, in many cases 40 to 60 per cent higher than the designed standards. In some areas, improper water regulation and drainage has also led to waterlogging. This is socially extremely wasteful in the context of the land water scarcity of the Indian economy. The issues involved in water conveyance are, therefore, of very high priority.

This phase is important for irrigation planning. This phase has to be seen not just as a question of regional studies. The association of the irrigation engineer with the wider disciplines of geography and land use, meteorology and ground water will from the beginning sensitize the irrigation planner to the diverse agro-climatic and agro-ecological environment in which the water system has to operate and this sensitivity is extremely important to planning the physical features of the irrigation structures.

There is the much vexed question of cropping patterns and the problems that arise when the 'design' cropping pattern does not materialise on the field or the 'sanction' for the crop(s) is violated in practice. The practice uptil now has been to consult agronomists and to fix these parameters into the design of the project canal capacities. The outcomes depended on the seriousness of the agronomist who gives the projection. In one particular case of a project which used to be singled out by Prime Minister Indira Gandhi for critical comment, (Tawa), the cropping pattern in the project area was taken from the Second Five-Year Plan of the particular State in which it was located. The project area is less than 5 per cent of this large paddy growing State. Paddy is totally unsuitable as a crop for the black soil in the command area. But somehow paddy was recommended as a crop for this project. Irrigation system designed for paddy of course created considerable havoc in the subsequent phases of development there. The Ken-Betwa repeats this.

Here again, there is considerable need to understand and quantify the behaviour of the Indian farmer in the agro-climatic regime being studied. Economists have for long worked on this problem and have developed 'acreage response models.' These essentially postulate that the acreage allocation of the farmer follows profit maximising behaviour and depends on rainfall and its distribution, irrigation, the agricultural technology available and relative prices. The use of these models for irrigation planning would mean that the farmer's behaviour is studied and statistically estimated through acreage allocation models for an area of the kind with due consideration of the agro-climatic characteristics of the region in the command. Such studies are done with the farmers in groups. This is being done now again in Dabhoi in SSP in Gujarat where farmers irrigation groups have been created. (See, A.B. Mandevia, in R. Subbiah, *et al.*, ed., pp. 35-37, for details).

For the different possibilities of water availability it would be possible to estimate alternative crop sets that the farmer would consider once irrigation is available to him. The model would be of an econometric variety as follows:

- We take a big irrigation command. With some simplification the same principles will apply to ground water development. The command has 'r' regions and 'c' crops. First take the obvious fact that regionalisation should be done with irri-gation characteristics namely rainfall and distribution, soil characteristics, water aquifer characteristics and the possible layout of delivery systems. Administrative and/or general economic development categories are not important. Assume that potential crops for the command are worked out with agronomic studies.

  For any region 'r', the potential crops 'c' are defined.

  Now we can first work out

  (i) Arc = f (R, I, Prc)

  Where,

  Arc = area in the rth region under crop 'c'

  R = rainfall (mean; variability can also be tried)

  I = Irrigated area

  Prc= Relative price of cth crop (or relative profitability can be tried).

  (i) can be estimated with distributed lags. Work out a relative price configuration for the future.

  Then for alternative levels of irrigated area, alternative crop sets combinations can be worked out (Arc/Ar). In the next step for each crop set, work out water requirements in peak periods Wr. Now with an aquifer model, limit surface water delivery such that water logging is ruled out. This gives alternative combinations of ground and surface water' delivery (totalling to W) which can be used potentially in each region.

- For each possible crop set in each region, using potential crop yield (Table 4.2) and cost of cultivation data for each crop, value added in agriculture can be worked out. Work out separately for wages (hired or imputed) and surplus.
- Given the estimates of costs of delivering water, maximise value added for all regions applying a social wage criterion which is derived after examining inter-regional irrigation possibilities. The important issues, therefore, are to use farmer's behaviour patterns to work out cropping possibilities; to allocate water subject to ecological constraints (water-logging and drainage parameters); and to explore in details the benefits arising from spreading water through the command rather than concentrating in a few areas to avoid the tail ender problem.
- Technological aspects like distribution systems with sufficient capacity, control (for regulation) and efficiency (to avoid losses) are as important as the use of appropriate economic mechanisms, e.g., pricing of water and incentives for groundwater use. An estimate of possible crop sets for one region of an irrigation command in which this method has been applied is given in Table 4.1. The method essentially reduces the theoretical cropping sets from infinite possibilities to a set of alternatives which need to be considered for irrigation design purposes. It also sensitises the water resources engineer to the possibilities of alternative agricultural development in the area in which he is working.
- Two further features can be noted, namely that systematic monitoring of ground water and the construction of water balance models at the level of an aquifer are in fact not very expensive propositions and can be followed through fairly easily with modern computers, once the basic information is available. Given preliminary estimates which can generally

Table 4.1

Projected Irrigated Crop Sets (Fraction of Cropped Area) for Region – 1

Sl.No.	*Crop1*	*Set 1*	*Set 2*	*Set 3*	*Set 4*	*Set 5*	*Set 6*	*Set 7*
1	Paddy	0.16	0.18	0.20	0.17	0.17	0.12	0.10
2	K *Jowar*	0.01	0.00	0.01	0.00	0.01	0.03	0.05
3	K *Bajra*	0.03	0.03	0.02	0.03	0.02	0.03	0.03
4	K Groundnut	0.00	0.00	0.00	0.00	0.00	0.00	0.00
5	Wheat	0.14	0.14	0.14	0.12	0.12	0.12	0.16
6	Vegetables	0.03	0.02	0 .02	0.04	0.04	0.04	0.02
7	R Pulses	0.04	0.04	0 .03	0.03	0.03	0.03	0.03
8	S *Bajra*	0.04	0.03	0 .02	0.03	0.03	0.04	0.02
9.	S Groundnut	0.01	0.00	0.00	0.02	0.02	0.05	0.00
10	R *Jowar*	0.00	0.04	0.05	0.04	0.04	0.05	0.05
11	Tobacco	0.09	0.09	0.05	0.04	0.09	0.05	0.02
12	Cotton	0.31	0.35	0.38	0.34	0.33	0.42	0.40
13	Sugar Cane	0.00	0.00	0.00	0.02	0.02	0.00	0.00
14	Perennials & Fruit Crop	0.02	0.02	0.02	0.02	0.03	0.02	0.02
15	Other Crop (including Lucerne)	0.06	0.60	0.05	0.05	0.05	0.05	0.05

be built with easily available bore hole data, soil characteristics and land use information, more refilled models can be developed as the investigation of soil characteristics and the ground water regime proceeds at micro levels.

- Now if information on alternative crop sets of the kind discussed earlier is available and the yield from irrigation can be estimated with the degree of precision indicated for a recent period and the costs of building conveyance systems to the field can be estimated accurately, a very precise estimate of the benefit of irrigation conveyance estimates can be derived (Table 4.2). Theoretical procedures are described earlier. Such calculations have in fact already been made for some advanced projects which are off the design board and we now have the techniques to replicate them on a much

Table 4.2

Irrigated and Unirrigated Crop Sets for Different Water Allocations for Region – 1

Upper Limit of Canal Utilisation ('000 Ha)	Peak Canal Water Demand ('000 Cusecs)	Peak G W Demand ('000 Cusecs)	Value of Objective Function (Rs. Crore)	Irrigation Intensity GIA CCA	Annual G W Utilisation ('000 hm)	Efficiency -60% Canal Utilisation ('000 ha) Kharif	Total	Irrigated Set Intensity Set No./Value	Unirrigated Set Inter Set No./Value
(1)	(2)	(3)	(4)	(5)	(6)	(7)	(8)	(9)	(10)
65	1.771	0.655	41.583	0.707	25.181	31.519	65.000	4/0.707	1/0.452
70	1.891	0.700	43.113	0.755	26.386	34.236	70.000	4/0.755	1/0.414
75	2.012	0.745	44.616	0.803	27.590	36.952	75.000	4/0.803	1/0.377
80	2.294	0.750	45.966	0.882	28.795	37.730	80.000	1/0.448 4/0.448	1/0.325
85	2.597	0.750	47.305	0.964	30.000	38.243	85.000	1/0.927 4/0.037	1/0.271
90	2.790	0.750	48.632	1.914	31.205	40.518	90.000	1/0.832 2/0.182	1/0.229
95	2.971	0.750	49.959	1.061	32.410	42.966	95.000	1/0.679 2/0.382	1/0.187
100	3.153	07.50	51.286	1.109	33.614	45.413	100.000	1/0.526 2/0.583	1/0.145
105	3.334	0.750	52.613	1.156	34.819	47.860	105.000	1/0.373 2/0.783	1/0.103
110	3.322	0.750	53.269	1.151	31.872	53.152	110.000	2/0.940 4/0.211	3/0.100
115	3.322	0.750	53.686	1.151	27.354	58.152	115.000	2/0.940 4/0.211	3/0.100
120	3.322	0.750	54.103	1.151	22.836	63.152	120.000	2/0.940 4/0.211	3/0.100
125	3.322	0.750	54.105	1.151	22.820	63.170	120.018	2/0.040	3/0.100

larger scale. With these kinds of calculations, if an explicit weight has to be given to generate more employment in some poverty-stricken region, this could be done in a quantifiable manner.

Irrigation projects have, therefore, now to be designed within the framework of a very detailed understanding of the agro-climatic and agro-economic regime for which they are being designed. It is possible to take into account the diverse features of the Indian agricultural economy to develop such designs as illustrated above. The paper by A.B. Mandevia cited above gives a counterfactual of how such a system designed for SSP was implemented and is at present being constructed. Pre-qualification of bidders for the Sardar Sarovar Canal Automation Project was advertised earlier this year (See for example, Times of India, Ahmedabad, February 20, 2005, p.9). This is for a Remote Monitoring and Control System for Canal Network (Canal Automation) interfacing with Supervisory Control and Data Aquisation (SCADA) System for the Pilot Project consisting of Narmada Main Canal Reach Km. 0.0 to 105.344, Vadodara Branch Canal (with Jambusar and Kundhela Branch Canals) and Sakarda Branch Canal Systems. To be implemented the delivery systems have to be designed in outline at the initial stages.

Conjunctive Use

The design for conjunctive use of surface and ground water has to be planned from the beginning. This requires emphasis on ground water investigation. Given the kinds of crop sets discussed above, the peak requirement of water can be worked out, for each crop set. Once the possibilities of safe withdrawal of ground water are known, the requirements of each crop set for the balance surface water can be estimated.

If this kind of planning is considered (See fig. on page 86) from the beginning and institutionalised either by promoting use of tubewells by the farmer or by integrating ground water into project, the possibilities of water-logging would be eliminated from the beginning.

Cropping Patterns

In spite of all the warnings the cropping patterns for irrigation design are of a traditional kind implying poor use of land and subsequent problems. No pastures, forests or suitable crops are there. The details are very disappointing. This was not done in the Betwa case and is not suggested in the TOR for the Interlinking Project.

The water use plan has no relation to soil type. The proposed cropping pattern is largely dependent on paddy as the series of Tables 8 reproduced from the Project Report show. This is also true of the requirements projected in the Ken, as shown below. The present rice yield is between 7 to 9 qtls/per hectare, lower than in the times of Harsha and wheat yields are less than 20 qtls per hectare. Yields for coarse cereals, (barley at above 15qtls/hec) oilseeds (8 to 12 qtls/hec) and pulses are good, but no lessons are learnt. The data given in the FR itself shows that at the Taluka levels there is a much lower level of irrigated area under paddy, but the mistake persists.

Table 4.3

Cropping Pattern in Upper Betwa Sub-Basin 81

No.	*Crop Type*	*Percentage Cropwise*
1.	Paddy	1.01
2.	Jowar	8.83
3.	Wheat	48.52
4.	Maize	1.58
5.	Pulses	26.98
6.	Sugarcane	0.20
7.	Vegetables and fruits	0.29
8.	Other food crops	1.20
9.	Oilseeds	4.04
10.	Fodder	7.35
	Total	100.00

Source: Feasibility Report of Ken-Betwa Link Project; National Water Development Agency, Ministry of Water Resources, Government of India.

Conclusion

Drainage, soil shaping, contour management, improvement and management of lower level canals, desilting of tanks, raising embankments, fish culture, market development, controlled grazing

Table 4.4

Proposed Cropping Pattern in Ken Command

Type of Crop	*Percentage of CCA*	*Area in Hectares*
Kharif Paddy	20	48,238
Jowar/Maize	8	19,295
Pulses	15	36,178
Oilseeds	12	28,943
Vegetables	5	12,059
Sub-total	60	1,44,713
Rabi Wheat (HYV)	40	96,475
Wheat (Local)	5	12,059
Gram	8	19,296
Linseed	5	12,059
Sunflower	5	12,059
Barseem	3	7,236
Masoor	8	19,295
Sub-total	74	1,78,479
Total	134	3,23,192

Source: Feasibility Report of Ken-Betwa Link Project; National Water Development Agency, Ministry of Water Resources, Government of India.

Table 4.5

Proposed Cropping Pattern and Cropwise Water Use in the Command of the Four Projects of Upper Betwa Basin

Type of Crop	*Per cent of Area*	*Cropwise water use in ham**			
		Neem-kheda	*Richhan*	*Kesari*	*Barari*
Kharif					
Paddy	25.00	270.41	9457.43	472.51	22343.91
Maize	2.50	1.73	60.40	3.02	142.69
Jowar	2.50	1.66	58.19	2.91	137.47
Vegetables	2.50	3.14	109.75	5.48	259.29
Pulses	6.25	3.95	138.10	6.90	326.28
Groundnut	5.00	3.95	138.47	6.92	327.15
Soya Bean	2.50	1.39	48.61	2.43	114.85
Fodder	1.25	0.68	23.94	1.20	56.56
Rabi					
Wheat	50.00	166.37	5818.82	290.72	13747.42
Vegetables	5.00	14.78	517.06	25.83	1221.61
Grams	12.50	132.75	1145.35	57.22	2705.98
Oilseeds	6.25	12.21	427.20	21.34	1009.30
Perennial					
Sugarcane	3.75	34.62	1210.90	60.50	2860.86
Total	125.00	547.64	19154.22	956.98	45253.3

* Water use for evaporation losses may be met out from ground water.

Source: Feasibility Report of Ken-Betwa Link Project; National Water Development Agency, Ministry of Water Resources, Government of India.

Table 4.6

Projected Water Requirements in Ken

Type	*Crop*	*Water Requirement (Mm3)*
Kharif	Paddy	313.06
	Jowar/maize	58.81
	Pulses	45.95
	Oilseeds	66.16
	Vegetables	36.76
	Subtotal	520.74
Rabi	Wheat (HYV)	588.11
	Wheat (local)	55.13
	Gram	83.32
	Linseed	39.82
	Sunflower	39.82
	Barseem	23.89
	Masoor	24.50
	Subtotal	854.60
	Total	1375.34

Source: Feasibility Report of Ken-Betwa Link Project; National Water Development Agency, Ministry of Water Resources, Government of India.

and so on can all be thought of as alternatives or add-ons. Conjunctive use has already been discussed. Involvement of farmers and villagers in the area and an understanding of the rural economy would be essential. But the basic flaw is the scant treatment given to the soil, the crop and animal husbandry of the region and its people. The project does not relate with the scarcities and potential of the region. On all these counts, the Ken-Betwa project needs to start all over again if the concept of the ILR is not to get a fatal blow.

Another way of looking at the system is to compare the delivery systems in the Ken-Betwa with state of the art irrigation projects in India today. Some of the large branches of the earlier reaches of the Sardar Sarovar Main Canal and the entire canal systems of the Ken-Betwa systems in the first of the interlinking of rivers project announced with considerable ceremony have a lot in common in the services they have to provide. They will serve black cotton and other difficult soil areas having rainfall, which is not scanty, but is erratic and the areas need drainage. To the Ministry of Water Resources of the country goes the credit of sharing with the world the Feasibility Report of the Ken-Betwa Project. This degree of transparency is unusual and deserves to be applauded and one hopes is the precursor of the Information Act. Sardar Sarovar was always a transparent project with the over 50 page Planning fcr Prosperity available in the late-eighties of the last century and published a little later. Those who criticised it paid it the compliment of never reading it, or ignoring its detail for that would bring them down to real time. But there the similarity between them ends.

The SSP Canals are exposing the critics who first said the canal was not possible, then that water would not flow in it and now that its original design will be impossible. Explaining the Remote Monitoring and Control System now under way, the specialists says it is essential because, water is scarce, it is needed in different ways, time is important and anyway we can solve the problem. They put it in expertese. "balancing of the demands and supply and no single solution is applicable to all the 13 different agro-climatic zones", "the very limited delta of 21" (53 cm) at the head regulator", "response time far more important considering the distance of source to the farthest delivery point (around 700 km.)", using the "conveyance system for storage also and thereby virtually bringing the reservoir closer to the delivery point. The Narmada Main Canal itself will provide a storage capacity of 209 MCM." In the planning phase they had studied the technologies in other countries notably France, the US and in North Africa and adopted it to Asiatic peasant agriculture. Now the SCP-GERSAR, a French outfit is supervising a 105 kms project for the main canal, 102 kms for the Vadodara Branch and some of the smaller branches. The country's top engineering companies have qualified for bidding for the job and the dream is on the way to reality of assured controlled water at the time of need. This kind of covering the last mile will not be possible in traditional designs which flood areas for example to grow paddy.

The Ken-Betwa Feasibility on internet is knowledge proof. The soil scientists did their job as good technocrats as we saw. Their message was that life is complex. Nowhere do they let you grow paddy. But the feasibility report is of the good old kind providing for 30 to 40 per cent of the area under paddy or flood irrigation for the water requirements of a very heavy kind totally unsuitable for the region. We never learn. Civil society protest is there including by some in the Group set up by the WWF. The Malwa region really needs assured water. The project must go back to the drawing board so that the concept of interlinking does not damage. Photo-ops with the great are not a substitute for hard work.

References

Alagh Y.K. (1988) Guidelines for Agroclimatic Planning: A Draft for Discussion, *Journal of Land Development*.

———. (1991) Indian Development Planning and Policy. WIDER Studies in Development Economics, Helsinki and Delhi, Vikas.

———. (1991) Sustainable Development, From Concept to Action: Techniques for Planners, UNCED.

———. and D. Buch, (1995), The SSP and Sustainable Development, in W. Fisher, Columbia University Seminars, below.

———. (2000), Water and Food Security in South Asia, Invited Plenary Lecture at World Water Forum and Ministerial Meeting, The Hague.

Alagh Y.K. (1991) Indian Development Planning and Policy. WIDER Studies in Development Economics, Helsinki and Delhi, Vikas

———. (1991) Sustainable Development, From Concept to Action: Techniques for Planners, UNCED.

———. (2000) Water and Food Security in South Asia, Invited Plenary Lecture at World Water Forum and Ministerial Meeting, Hague.

———. (2002) Water Security: An Interdisciplinary Research and Policy Agenda, in Henk J. van Rinsum and Arie de Ruijter, More on MOST, pp. 10-30.

———. (1991) Sustainable Development: From Concept to Action, in UNCED and UNDP

———. (2000), Sustainable Development Policies for Developing Countries,Keynote Address, UNU/IAS-OECD Conference on Sustainable Development, Tokyo.

Alexandratos, N., (Ed.), (1995) World Agriculture Towards 2010, An FAO Study, Chichister, Wiley.

Colombo, U., (1986) New Technology, Bulletin of Social Information, 2, pp. 232-237.

F.A.O. (1993) **Agriculture Towards 2010**, Rome, F.A.O., for final published version see, N. Alexandratos

Falkenmark, M., J. Lundqvist and C. Widstrand (1989) Aspect of Vulnerability in Semi-Arid Development, NaturalResource Forum.

Falkenmark, M., (1991) Urgent Need for Water Perspective, Water International, Vol. 16, pp. 229-240.

Fisher, W., (1995) Towards Sustainable Development, Columbia University Seminar Series, New York, Sharpe.

Government of India (1989), Planning Commission, P.P.D., **AgroClimatic Planning : An Overview**, New Delhi, Planning Commission (authorship, Y.K. Alagh, *et al.*)

Nasution, L., (1993) Agricultural Regionalistion of Indonesia, Bogor, Agricultural, Research Centre.

Sachs, I., (1991) Background Paper for the Hague Declaration, extensively reprinted in Nature and by RIS.

Sadasyuk, G., and P. Sengupta (1968) Economic Regionalisation of India: Problems and Approaches, New Delhi, Census of India, Monograph No. 8 of Census 1960.

Sonntag, N., (1996), Adaptive Management Policy Exercises: Two Methods for Integrated Resource Management, in MRC., above, p. 27.

5

The Feasibility Report of Ken-Betwa Link Project: An Analysis of Assumptions and Methodology

KANCHAN CHOPRA

Introduction

The ILR project in its entirety, with a cost estimate of Rs. 5,60,000 crore elicits many different kinds of reactions, even before the constituents are identified and their feasibility studied. No single overarching view to the different links can be taken. Each link needs to be studied using "state-of-the-art" methodology before arriving at any conclusion. In this paper, the Feasibility Report of Ken-Betwa Link Project (FR) is examined with a view to:

- Summarising what it achieves, in particular in the context of economic and social issues.
- Analysing the assumptions with regard to future development of the region which it assumes.
- Examining areas of incomplete analysis.
- And finally, looking at international "best practice" both in terms of tools and techniques for analysis and in realising the limits to knowledge.

Feasibility Report of Ken-Betwa Link

The proposed Ken-Betwa link envisages the construction of

- A dam at Daudhan on the river Ken with a gross storage capacity of 2775 Mm^3.
- Two power houses with installed capacity of 3x20MW and 2x 6 MW respectively.

- Diversion of 659 Mm^3 of water (using existing outlet of Barwa Sagar) to Betwa basin upstream of Parichha weir to be utilised by way of substitution to provide annual irrigation to 1.27 lakh hectare of drought-prone area of upper Betwa basin. However, total diversion of water from the Ken basin will be 1020 Mm^3, after consideration of in basin demands and downstream commitments.
- Annual irrigation to an area of 47,000 hectare enroute of the Ken-Betwa link.
- Annual irrigation of 3.23 lakh hectares (CCA of 2.41 lakh hectares) under the Ken Multipurpose project" earlier proposed by the Madhya Pradesh Government.
- Provision of 11.75 Mm3 for drinking water supply to villages and towns enroute of the link canal.

The report is divided into thirteen chapters: four of these deal with hydrology, design features and related issues; one is on inter-state aspects, four with irrigation planning, and cost and financial aspects. In this paper, we examine Chapters 8, 9, 11, 12 and 13 for the assumptions on which they are based in the light of international practice in this context.

Assumptions Regarding Future Development of the Region and Possible Alternative Scenarios: Project Approach or Regional Development Approach

The question that the Ken-Betwa Feasibility Report, like any other project report needs first to ask is: What is the future development scenario for the region, how does the proposed project fit into that scenario taking due account of the human, land and water resources of the region? In this context, a few key decision points need to be taken into account in planning and project development.

The first of these is: needs assessment or validation of the needs for water and energy services and identification of the preferred development plan among the full range of options. Only after this stage do the succeeding stages such as project preparation, implementation and operation follow. A clear set of development objectives needs to be in place. This can follow from a scenario development exercise for the region. The use of scenario analysis to examine alternative futures is now a part of planning now. Complex issues involving socio-ecological systems are addressed using this approach.

Scenarios are "plausible, challenging, and relevant stories about how the future might unfold that can be told in both words and numbers. Scenarios are not forecasts, projections, or predictions." At best, scenarios might be considered conditional projections in that particular outcomes "reflect different assumptions about how current trends will unfold, how critical uncertainties will play out, and what new factors will come into play" (UNEP 2002, Robinson 2003). It is important to note that scenarios are not merely alternative runs of a model or a sensitivity analysis, although these can be important for looking at uncertainty within scenarios.

If done properly, particularly when formal quantitative models are used, all of the underlying assumptions are made explicit in a scenario. Certain assumptions take precedence, however. These represent the primary axes along which the scenarios differ and are generally related to contextual issues. These fundamental differences provide the 'logic' behind the individual scenarios and the scenario exercise as a whole (Schwartz 1996). The development of such scenarios in the context of the Ken-Betwa link can, for example be as follows:

- Possibility of accrual of livelihoods from sources other than agriculture, including ecosystem service related livelihoods.
- Higher levels of literacy of the region with high rates of developmental migration.
- Suitability or otherwise of the soil in the region for irrigated agriculture as the primary source of livelihood.

Such a scenario analysis with regional development as its focus places the project being considered in the perspective of object of the ILR project, stated to be "to provide national water security and alleviate poverty with a broad measure of regional and social equity."

In the context of future scenarios of this region's development, the Planning Commission in its study on Agroclimatic planning (Planning Commission, 1989) describes this area of 37 million hectares a part of the Zone 8 called the Central Plateaus and Hills. It identifies that it needs soil and water conservation by integrated watershed development.[1] How does this match with the assumptions made in the FR. The Feasibility Report, however, in particular Chapters 8 and 9 focus on irrigation planning and command area development. The implicit assumption is that

- the region, its agro-ecology is suited to irrigated framing; and
- an increase in the accrual of income from agriculture is the main source of development and poverty alleviation.

A detailed analysis of the compatibility of more intensive watershed development programmes in the upper catchment with water availability for the project site needs to be undertaken. A water balance exercise for the region, both with and without extension of soil and water conservation in the upper catchment needs to be undertaken.

Further, the primacy of agriculture as the basis for poverty alleviation and development is established based on data and a field study conducted in 1991-92. In the period since then, the options available to people in the region have, presumably, undergone a change and may warrant a study of different scenarios for the accrual of livelihoods. With increased literacy and expansion of alternative opportunities, for example, other service sectors could be conceived of as contributing to incomes and poverty alleviation. While it is expected that the Detailed Project Report will examine these aspects, the Terms of Reference of the DPR[2] continues to assume that irrigated agriculture comprises the only option for development in the region.

Costs, Revenues and C-B Analysis as the Decision Support Framework

The FR uses Cost-Benefit Analysis as the Decision Support Framework. This is standard practice in India and guidelines exist (See for instance, Government of India, 1983). A number of versions of the same exist such as financial analysis, economic analysis and so on. It seems as though financial analysis alone has been used. Even in implementing that, existing reality must be taken into account. Chapter 11 of the FR, for instance, is entitled "Environmental and

1. See also Alagh's paper on the Ken-Betwa link.
2. See NWDA, "Terms of Reference of the Detailed Project Report Interlinking of Rivers."

Ecological Aspects of the Project." While the chapter does list the species found in the area and the loss of forests to submergence, it fails to make the link of these with human well-being and the provision of livelihoods. It ignores the large and growing literature on how different kinds of ecosystems contribute to livelihoods and human well-being. It also ignores existing legal stipulations with regard to conversion of forestland to other uses.

Some details given below raise concerns that are worth taking note of:

- Provision is made for compensation for government/private land and forest lands coming under submergence. It is not clear at what rates these provisions are made. They may need upward revision to conform to present day norms. Note for instance that 6,400 hectares (74 per cent of the submerged area) is under forests. As per Supreme Court ruling of November 2002, the net present value of forest land diverted must be recovered at the rate of Rs. 5.80 lakh per hectare to Rs. 9.20 lakh per hectare.
- Of the forest area to be submerged, 45 square kilometers is part of Panna National Park and cannot be converted to other uses as per law. Apart from its ecological value, these areas provide livelihoods from tourism and related indirect incomes. Chapter 11 states, "the impact on wild life of the park will be almost nil as the area coming under submergence is only about 9 per cent of the park area and wild life has its characteristic of moving to the interior forest areas!" Even if this is true, legal stipulations exist and there is a cost attached to following them. Financial analysis of the project must take account of these. Further, only a part of these costs may be treated as transfer payment in the economic analysis. Cost estimates under 'environment and ecology' are again abysmally low at Rs. 16.18 lakh, supposedly to include soil conservation measures, compensatory afforestation and a number of other line-items.
- Additionally, plantation has been allotted Rs. 3.62 lakh. Similar items exists under other units as well. This is presumably plantation in residential colonies etc.

Other concerns with respect to costs (Chapter 12) are:

- Where is the cost of R&R (Resettlement and Rehabilitation) provided for? It is not clear.
- It is mentioned in Section 9.12 that "land is generally undulating. Hence, the canal distribution system may have to be aligned accordingly?" Does this imply a higher cost? Is it factored into the cost?

Sources of revenue in Ken-Betwa Link project as stated in the Report are:

- Irrigation Water rates: The rates used in the Feasibility Report are not stated. If the norms of the Vaidyanathan Committee are followed, working expenses on the canal, dam and other works need to be covered. These are Rs. 180 per hectare of annual irrigation.
- Betterment Levy for 20 years after the irrigation begins, at stated rates.
- Revenue from Power.
- Irrigation cess.

The Feasibility Report ends by saying:

"The B:C ratio for the project as a whole (including power) has been worked out and it is obtained as 1.87. The net annual benefits from the irrigation component are obtained as 449.79 crores. The IRR (Internal Rate of Return) from the irrigation component has been obtained as 13 per cent."

There seems to be an implicit assumption that the B:C Ratio and the internal rate of return shall comprise the decision criteria. The fundamental decision steps underpinning cost-benefit analysis are summarised in Hanley *et al.* (1997), among other places:

- Derive estimates of costs and benefits,
- Rank initiatives from high to low in terms of net benefits, and
- Pursue as many initiatives (with positive net benefits) as possible within resource constraints.

It is important to recognise from the outset that the cost-benefit approach to decision-making ignores the distribution of costs and benefits — an omission that can have political and social implications. Programmes or projects are judged to be attractive as long as total (expected) benefits exceed total (expected) costs regardless of who bears the cost and who enjoys the benefit.

Choosing the correct discount rate is also an enormous issue when costs and benefits extend into the future because discounting can render long-term future effects almost irrelevant in the calculation of discounted net benefit. Short-term calculations are less problematic. Markets are driven by private agents who discount the future at the return to private capital — that is, the opportunity cost of financial capital.

Expected net benefits can be the basis of decision rankings if analysts can assess probabilities (even based on subjective judgments) across a range of outcomes that accommodate the full range of possibilities, including the possibility. If analysts cannot assess probabilities, they can still assess robustness – the range of possible outcomes for which net benefits are positive and perhaps identify critical thresholds for critical sources of uncertainty along which net benefits turn from positive to negative. We assess the state of knowledge with respect to some of these decision criteria below.[3]

By way of extensions of the analysis, it is suggested that benefit cost analysis basically amounts to calculating the value of water delivered for agriculture at specific locations at specific times. It may be worth working this value of delivered water explicitly. It may help in assessing the scarcity value of water in different basins, which is quite crucial when inter-basin transfers are being considered. It may also help in working out the opportunity cost of water when uses other than agriculture are being considered.

The examination of conjunctive use and the impact on the water basin as a whole requires us to step outside the project benefit-cost framework and look at the impact of the project on the entire watershed (in the source and the destination area when inter-basin transfers are being considered). As argued earlier, such eco-system

3. This section draws heavily from Millennium Ecosystem Assessment, Report of the Responses Group (forthcoming). Chapters 4, ...)

based area analysis needs to be combined with more traditional project analysis.

Complementing Decision Support Frameworks and their Role in ILR Projects

The state of knowledge with respect to decision criteria for projects with and without certainty has gone a long way and a range of tools are now available which need to be used in project evaluation, in particular for projects which impact the environment.

Every framework designed to support decisions about response options must be able to accommodate uncertainty in its application; the diversity of available frameworks is outlined in the MEA framework volume (MEA 2003, p.196) with the advantages and disadvantages of each identified.

Some frameworks, which can be used with profit in providing information to enable decision makers to arrive at policy decisions, are listed below.

Multicriteria Analysis (MCA)

Issues derived from conflicting interests in the management of a given resource pose particular problems, especially when distributional implications are to be considered. Multicriteria analysis (MCA) and its variants are often the formal framework of analysis used to decide between various response options under these circumstances. Formal multicriteria analysis can trace its roots to Pareto at the end of the 19th century. In the 1970s, the development of multi-objective maximisation methods permitted widespread application of quantitative MCA, especially in the management of water resources and (more recently given the advent of GIS) land-use planning. Multicriteria analysis depends on completing a number of concrete steps:

- Identifying objectives;
- Identifying options for reaching these objectives;
- Identifying evaluation criteria;
- Analysing options against those criteria;
- Making choices based on those analyses; and

- Evaluation and feedback.

MCA's explicit recognition of a multiplicity of objectives and evaluation criteria gives it a potential advantage over economic paradigms based on cost, benefit, and efficiency in identifying sources of vulnerability to uncertainty and even to ignorance. Using the framework outlined above, it is possible to discover uncertainty in objectives, unexplored options, incomplete evaluation criteria, ignorance about system properties, and volatile rules of choice.

Precautionary Principle and "Safe Stopping Rules"

The precautionary principle is an approach used by policymakers in which they consider taking action to protect a population from potential hazards with serious or irreversible threats to health or the environment before there is strong proof that harm will occur. In essence, the precautionary principle prescribes how to bring scientific uncertainty into the decision-making process by explicitly formalising precaution and bringing it to the forefront of the deliberations (Marchant 2003). It posits that significant actions (ranging from doing nothing to banning a potentially harmful substance or activity, for instance) may be justified when the degree of possible harm is large and irreversible. Many factors influence these deliberations, including an assessment of the possible severity of the potential harm and the degree of scientific uncertainty associated with that assessment.

The application of the precautionary principle to environmental hazards and their uncertainties began with the Swedish Environmental Protection Act of 1969, with further elaboration in the German Clean Air Act of 1974 and the 1985 report on the Clean Air Act (Kheifets *et al.* 2001, Boehmer-Christiansen 1994, EEA 2001). Since the 1970s, the precautionary principle has been incorporated into over a dozen international environmental agreements, expressly incorporated into the legal framework of the European Union, and adopted into the domestic laws of numerous nations (Marchant 2003). This principle featured in the 1992 Rio Declaration on Environment and Development as Principle 15 (UN 1993): "In order to protect the environment, the precautionary approach shall be widely applied by States according to their capabilities. Where there are threats of serious or irreversible damage, lack of full scientific certainty shall not be used as a reason for postponing cost-effective measures to prevent environmental degradation."

In this definition, the precautionary principle is 'risk-oriented' in that it requires evaluations of risk that include cost and benefit considerations (Tamburlini and Ebi, 2002). It is clearly intended for use in drafting provisional responses to potentially serious health or environmental threats until adequate data are available for more scientifically based responses. It also can be applied when there may be undue delay in the regulatory process. It is easy to see that the interlinking of rivers project in India may be confronted with many situations of incomplete knowledge and would need to contend with this principle. In some cases, the conclusion may be that the precautionary principles would not come into play. However, TORs for Detailed Project Reports would need to take cognizance of it.

Vulnerability Analysis

Methods of vulnerability assessment have been developed over the past several decades in addressing natural hazards, food security, poverty analysis, sustainable livelihoods, and related fields. These assessments have helped to determine the baseline characteristics of those individuals, groups, or ecosystems that are sensitive to changes and shocks in the system. Vulnerability assessment (VA) can identify both general and specific vulnerabilities that enable targeted intervention and can guide future development projects. In identifying the most vulnerable systems or groups, measures to increase social resilience and ecosystem productivity can be prioritised.

Vulnerability can be defined as the degree to which "an exposure unit is susceptible to harm due to exposure to a perturbation or stress and the ability (or lack thereof) of the exposure unit to cope, recover, or fundamentally adapt (become a new system or become extinct)" (Kasperson and Kasperson, 2001).

An alternative approach focuses on the agents of interest and their underlying internal characteristic (Brooks 2003). This can be termed "social vulnerability." Studies in this vein explicitly consider livelihood elements (such as access to information and resources) as well as the broader socio-economic environment (such as factors that determine the ability of the system and agents within the system) to cope with their existing situation and to respond to potential impacts (Adger 1999, Blaikie *et al.*, 1994; *see* also Turner *et al.*, 2003b).

VAs are beginning to move beyond static snapshots at a particular time and place to assessments that depict cumulative and long-term vulnerability at a variety of spatial scales. It has been recognised that vulnerability cannot be viewed as a static phenomena. Vulnerability is, rather, part of a dynamic process (Leichenko and O'Brien 2002). This recognition has implications for the level of uncertainty associated with vulnerability assessments because dynamic analyses require information about how processes and characteristics change over time.

Wherever project impacted people are disadvantaged, vulnerability assessments are likely to play a critical role.

Risk Assessment

The classic risk assessment approach to, for example, evaluating health or ecological risks, adopts a four-step risk paradigm:

- Identify the hazard (could a particular agent or activity harm humans, animals, or plants?);
- Assess the exposure-response relationship (to what degree can exposure to a hazard cause a response that could be harmful?);
- Assess the level of exposure (to what degree are humans, animals, or plants exposed to the hazard); and
- Characterise the associate risk (a reflection of the probability exposure times the associated consequences).

Each step involves a policy judgment; for example, the choice of one dose-response model over another is a "science-policy" choice (NRC 1994, 1996; US Presidential Commission 1997). The limitations of risk assessment should be recognised and understood. The underlying assumptions may, first of all, limit its applicability to complex environmental problems (Bernard and Ebi 2001). The assumption that a defined exposure to a specific agent causes a specific adverse outcome for identifiable exposed populations can, in particular, be questioned in many contexts. A health outcome may be distinctive and the association between immediate cause and its impact can be fairly clearly determined.

In the context of ILR projects, management of disasters such as earthquakes and risk of exposure to diseases such as malaria can be analysed using this approach.

Applicability of Decision Support Methods and Frameworks

The following table indicates how different decision support frameworks assist the decision maker. Each provides a different kind of information on different dimensions of the project. Put together, they provide a range of information which can then be used in decision-making.

Table 5.1

Applicability of Decision Support Methods and Frameworks

Method	*Efficiency*	*Equity*	*Consideration of Thresholds*	*Uncertainty*	*Scale*
Cost Benefit Analysis	+	-	-	A	X
Risk Assessment	+	-	*	E	X
Multicriteria Analysis	*	+	+	A	N and M
Precautionary Principle	+	+	*	E	X
Vulnerability Analysis	+	+	*	A	N and M

Interpreting the efficiency, equity and thresholds columns: (*) designates direct applicability by design; (+) designates possible applicability with modification; and (-) designates weak but not impossible applicability with considerable effort. Interpreting the uncertainty column: (A) designates a method that has been modified to accommodate uncertainty and (E) designates a method that has been designed explicitly to handle uncertainty. Interpreting the scale column: M designates primary applicability at a micro scale; N designates applicability up to a national scale; R designates applicability to a regional or sector scale; G designates applicability up to a global scale; and X designates applicability at all scales.

Feasibility studies for new projects such as the Ken-Betwa link need to take account of dimensions of the above developments which are relevant.

Characteristics of Decision-Making Processes

What process do we adopt in arriving at decisions with respect to large public investments? International best practice tends now to give due consideration to the processes that result in decisions as much as to the criteria used. Analysis of a large number of projects in different parts of the world illustrates that adaptive and participatory processes of decision-making yield more lasting results, even though they are more resource intensive and time consuming. The following elements of decision-making processes tend to improve the decisions reached and their outcomes.

- Use the best available information.
- Ensure transparency and the effective participation of all concerned and affected stakeholders.
- Recognise that not all important values at stake can be quantified and, thus, quantification can provide a false objectivity in decision-processes that have significant subjective elements.
- Strive for efficiency, but not at the expense of effectiveness.
- Strive for equity and reduce vulnerability.
- Ensure accountability and provide for monitoring and evaluation which includes stakeholder feedback.
- Consider cross-scale effects.

In the context of water development projects in particular, the understanding is that project short-listing, selection, execution and monitoring needs to be based on recognising entitlements to water and pre-existing water rights, in particular of vulnerable groups of people.

Lessons for Feasibility Studies and TORs for Detailed Project Reports

The Feasibility Report of the Ken-Betwa falls short of best practice on a large number of accounts. It does not contain any reference to use of techniques or methods of analysis other than cost-benefit analysis. It assumes, without projecting future scenarios for the region that irrigation based agriculture is the only development path available to the region. A more comprehensive project report shall need to take note of the state of knowledge with regard to these areas of analysis. That alone can provide an effective and useful basis for decision-making on the Ken-Betwa link or, for that matter on any other projects envisaging Interlinking of Rivers in India.

The TOR for preparation of Detailed Project Reports (DPR) leaves scope for some of the concerns mentioned above. There exist 27 points mentioned in 3.2 under the scope for preparation of the DPR. The most glaring omissions are with regard to:

- The scenarios for regional development and how the project fits into these projected scenarios.
- The excessive stress on numerical and modelling aspects without a spelling out of modelling of other non-project modes of use of water resources of the region
- The lack of reference to any decision support framework other than benefit cost ratio, whether as economic analysis or as returns to the state exchequer.

References

Adger, W.N. and C. Luttrell (2000) "Property Rights and the Utilisation of Wetlands," *Ecological Economics*, 35, 78-91.

Bernard, S.M. and K.L. Ebi (2001) Comments on the Process and Product of the Health Impacts Assessment Component of the United States National Assessment of the Potential Consequences of Climate Variability and Change, *Environmental Health Perspectives*, 109(Suppl 2), 177-184.

Blaikie, P., T. Cannon, I. Davis, and B. Wisner (1994) *At Risk: People's Vulnerability and Disasters*, Routledge, London.

Boehmer-Christiansen, S. (1994) "The Precautionary Principle in Germany: Enabling Government." in *Interpreting the Precautionary Principle*, T. O'Riordan, and J. Cameron (eds.), Cameron and May, London.

Brooks, N. (2003) *Vulnerability, Risk and Adaptation: A Conceptual Framework*, Tyndall Centre for Climate Change Research, Working Paper 38.

EEA, (2001) *Late Lessons for Early Warnings: The Precautionary Principle, 1896-2000*, European Environment Agency, Copenhagen.

Government of India, Central Water Commission (1983) Report of the Committee to Review the Existing Criteria for Working out the Benefit-Cost Ratio (Nitin Desai Committee Report).

Hanley, N., J. Shogren, and B. White (1997) *Environmental Economics in Theory and Practice*, Oxford University Press, New York.

Kasperson, J. and R. Kasperson (eds), (2001) *Global Environmental Risk*, United Nations University Press/EarthScan, London.

Kheifets, L. I., G. L. Hester, and G. L. Banerjee (2001) "The Precautionary Principle and EMF: Implementation and Evaluation," *Journal of Risk Research*, 4, 113-125.

Leichenko, R. and K. O'Brien (2002) "The Dynamics of Rural Vulnerability to Global Change," *Mitigation and Adaptation Strategies for Global Change*, 7, 1-18.

Marchant G.E. (2003) "From General Policy to Legal Rule: Aspirations and Imitations of the Precautionary Principle," *Environmental Health Perspectives*, 111, 1799-1803.

Millennium Ecosystem Assessment (2003) Ecosystems and Human Well-being: A Framework for Assessment, Island Press, Washington DC.

National Water Development Agency (first circulated in 1996), Feasibility Report of Ken-Betwa Link.

National Water Development Agency, Terms of Reference for Preparation of Detailed Project Report: Interlinking of Rivers.

National Research Council CoRAoHAP, Board on Environmental Studies and Toxicology, Commission on Life Sciences, 1994: *Science and judgment in risk assessment,* Washington, DC: National Academy Press.

National Research Council CoRC, (1996) *Understanding Risk: Informing Decisions in a Democratic Society,* Washington, DC: National Academy Press.

Planning Commission,India (1989) Agro-climatic Zones: Profiles and Issues, ARPU Working Paper No. 2, Ahmedabad, Agroclimatic Regional Planning Unit.

Robinson, J., (2003) "Future Subjunctive: Backcasting as Social Learning," *Futures,* 35, 839-856.

Schwartz, P., (1996) *The Art of the Long View: Planning for the Future in an Uncertain World,* Currency Press, New York, Currency.

Tamburlini, G., and K.L. Ebi, (2002) "Searching for Evidence, Dealing with Uncertainties, and Promoting Participatory Risk-management," in Children's health and environment: A review of evidence, Tamburlini G., O.S. von Ehrenstein, R. Bertollini, editors. A Joint Report from the European Environment Agency and the WHO Regional Office for Europe, EEA, Copenhagen, 199-206.

Turner, II, B., Kasperson, R., Matson, P., McCarthy, J., Corell, R., Christensen, L., Eckley, N., Kasperson, J., Luers, A., Martello, M., Polsky, C., Pulsipher, A., Schiller, A., Selin, H., and Tyler, N., (2003b), Illustrating the coupled human–environment system for vulnerability analysis: three case studies, *Proceedings of the National Academy of Sciences US,* 100, 8080-8085.

United Nations (UN), (1993) *Agenda 21: The UN Programme of Action from Rio,* United Nations, New York.

United Nations Environment Programme (UNEP), (2002) *Global Environmental Outlook 3: Past, Present, and Future Perspectives,* EarthScan, London.

US Presidential/Congressional Commission on Risk Assessment and Risk Management, (1997) *Risk Assessment and Risk Management in Regulatory Decision-Making,* Washington (DC).

The author is thankful to the members of the National Civil Society Committee on Interlinking Rivers for comments after the meeting at Orchha, March 13-14, 2005.

6

Ken-Betwa Link: Why it Won't Click?

HIMANSHU THAKKAR
BIPIN CHANDRA CHATURVEDI

Ken-Betwa Link Project (KBLP) is one of the links among 30 River links proposed by the Government of India's National Water Development Agency (NWDA), involving MP and UP in the Bundelkhand region. A 73.8 m. high Greater Gangau Dam (GGD) is proposed on Ken River near Daudhan village, on the border of Chhatarpur-Panna districts in Madhya Pradesh, 2.5 km. upstream from the existing Gangau Weir. The water is to be transferred to Betwa river through a 231.45 km. long concrete lined link canal which is to drop water upstream of the existing Barwasagar reservoir in Jhansi district in UP. The project is supposed to strengthen the Paricha barrage canal system. Two power projects (60 MW and 12 MW) are also planned on the left bank of the proposed dam.

KBLP at 1st Meeting of Task Force: At the very first meeting of the Task Force on Interlinking of Rivers (ILR) held on Feb. 10, 2003 it was decided that UP and MP may finalise and confirm their master plans pertaining to water resources development within a period of one and two months respectively so as to proceed further regarding KNLP (GoI, MoWR, Monthly Summary for the Cabinet for Feb., 2003).

However, it took much longer for the agreement between MP and UP to materialise. On August 25, 2005, in the presence of Prime Minister Dr. Manmohan Singh, UP and MP Chief Ministers signed the Memorandum of Understanding for taking up Detailed Project Report (DPR). The DPR is to cost Rs. 40 crore and Rs. 6 crore has been allotted in the 2005-06 budget. As per the latest available information, the Central Water Commission or Engineers India Limited was to be given the work of preparing the DPR for the Link.

Task Force

According to the Chairman of the ILR Task Force the Ken is to be linked with the Betwa through a canal to replace the 659 MCM (Million Cubic Meters) of Betwa waters reaching Paricha weir by equivalent supply of water from Ken. This would enable annual irrigation of about 0.127 M Ha in Raisen and Vidisha districts of MP in the upper Betwa basin. The Daudhan reservoir link canal is to divert about 1020 MCM of water from the Ken basin. Out of this flow, about 312 MCM is to be used for enroute annual irrigation of about 40000 Ha in Chhatarpur and Tikamgarh districts in MP and 7000 Ha in Hamirpur and Mahoba districts in UP. The link is also to provide 11.75 MCM water for drinking water supply in these enroute districts. This apart, water from the Daudhan reservoir is to be utilised to irrigate about 0.323 M Ha in Panna and Chhatarpur in MP (water used will be 1375 MCM) and about 0.148 M Ha in Banda district in UP [This last claim sounds rather strange, as this is not mentioned in the official National Council for Applied Economic Research report on Socio Economic and Environmental aspects of six links, done in 1994 for National Water Development Agency.]

Ken River Basin

Rising from the north-west slopes of the Kaimur hills in Jabalpur district at an elevation of about 500 m. above the mean sea level, the average annual flow of the river is 11300 MCM. It is 427 km. long up to its point of confluence with the Yamuna near village Chilla in the Banda district of UP. The river basin lies between north latitudes 23°20' and 25°20' and east longitudes 78°30' and 80°36'. The Ken basin covers the area of Jabalpur, Sagar, Damoh, Panna, Satna, Chhatarpur and Raisen districts of MP and Hamirpur and Banda districts of UP. It is bounded by Vindhya ranges in south, Betwa basin is in the west, free catchment of the Yamuna in east and the river Yamuna towards the north. The catchment area of Ken basin is 28224 km^2.

Tributaries: Chandrawal, Urmil, Shiam, Shihu, Karoran, Kel, Bichhui, Gawain, Siamri, Banne, Khuraran, Kutni, Lohruk, Kusar, Kail, Sonar (Tr-Bewas, Bamner, Kopra), Vyarma, Aloni, Vearma.

Some Existing Major Projects on Ken River

SN	*1*	*2*	*3*
Project	Gangau Weir	Rangwan	Bariyarpur
Year of Completion	1915	1957	1905
Height Above Lowest Foundation, m.	16.15	36	27
Gross Capacity of Res. (MCM)	119.43	163.62	12.59
Purpose	Irrigation	Irrigation	Irrigation
Max. Discharge Cap. of Spillways, Cumecs	13700	2410	14275
Benefit 1000 Ha	—	37.64	—
Catchment area sq. km.	18637	828	20760

Dams on Tributaries of Ken River

Dam	*Year*	*River*	*Ht (m)*	*Gross Storage Capacity MCM*
Chandrawal	1973	Chandrawal	10.97	34.71
Urmil	1994	Urmil	18.24	116.60

Gangau Weir: Gangau Weir in Chhatarpur district is a large water body formed by damming the river Ken at Nonapanji within this district. Gangau reservoir serves as a feeder dam to the Bariyarpur Weir from where numerous canals have been taken out to Panna (MP) and Banda (UP) districts of Bundelkhand. At present Left Bariyarpur Canal is under construction. UP objects to the construction of Left Bariyarpur Canal until the completion of Greater Gangau Dam. According to UP, it is impossible to supply water to Left Bariyarpur Canal from Bariyarpur Weir, as there is already shortage of water in Ken Canal System. In view of MP, Left Bariyarpur Canal is constructed as per 1972 agreement and remodelling of Ken Canal was allowed with the condition of construction of Left Bariyarpur Canal. MP has already spent Rs. 500 Million on canal construction. The issue remains undecided even after discussion in meeting of Central Regional Board on 18.02.2000.

Rangwan: The Rangwan dam (capacity 152.14 MCM) is built by UP on Banne river (Tributary of Ken) in Chhatarpur district, which is also used to feed Bariyarpur Weir as well as provide irrigation in Chhatarpur and Panna districts. MP has completed canal of length 25 km., as per agreement made in 1972.

There is a dispute between MP and UP on this project. MP has demanded for control over

i) Regulation of gates of Rangwan dam,

ii) Inspection houses constructed at Gangau, and

iii) Link road between Rangwan and Dhamari.

MP is demanding 55.53 MCM of water in *Rabi* season, which was agreed in 1972 for *Kharif* season. It was agreed upon that, available water in Rangwan Reservoir will be distributed between UP & MP in ratio of 36:15, But by providing 55.53 MCM of water to MP in *Rabi* season, the ratio will become 16:35, which is against bilateral agreement, as per UP. Also it will badly affect irrigation in UP.

Urmil Dam: Urmil Dam is situated in village Shamshera, Dist. Mahoba on Urmil River. Completed in 1994, the Urmil Dam is to provide Annual irrigation to 4 769 Ha from CCA of 6800 Ha of Dist Mahoba and Chhatarpur (MP) through main canal of 29.4 km. and distribution system 18.75 km. It provides 1.7 MCM water for drinking water supply. The 18.34 m. high dam has live storage capacity of 111.5 MCM.

Bariyarpur Weir: Bariyarpur Weir is built at Ajaigarh in Panna district in MP in 1905 by UP. It was built essentially to irrigate the agricultural area of Banda through Ken-canal system but since the reservoir was located in this district, the benefit of recharging of groundwater reserve naturally goes to Panna district. Bariyarpur Weir is supposed to provide irrigation to 0.23 M Ha in Banda and Chitrakoot (UP) and Chhatarpur districts (MP) through main canal of 59.34 km. length and distribution system 960.56 km. The annual gross irrigation is 65950 Ha. The total area submerged due to project is 3078 Ha, including 119 Ha forest land. At least 3000 people are known to have been displaced due to this project.

Ken Canal System, having 2500 cusecs capacity (after remodelling), originates from right bank of Bariyarpur Weir to irrigate areas in Panna (MP) and Banda (UP).

Left Bariyarpur Canal: According to agreement of 1972 & 1977, it was proposed to construct Left Bariyarpur Canal (1385 cusecs capacity & 59.38 km. length) and Greater Gangau Dam by MP. UP agreed for construction with the condition that, both the projects should be constructed at the same time.

Augasi Pump Canal: Augasi Pump Canal augments water to tail portion of Ken Canal System; Sanda Distributory and Adhayu Minor. It provides irrigation to CCA of 13360 Ha. It has 68 km. distribution system including 3.34 km. long Augasi main canal. Constructed in 1981-82, the canal has capacity to transfer 150 cusecs of water from Yamuna River to Ken canal system. However, since only 50 cusecs was available, reducing the *Kharif* Irrigation from planned 6012 Ha to 306 Ha and *Rabi* irrigation from 6680 Ha to 2772 Ha, the scheme is now being modernised to increase the utilisation.

Chandrawal Dam: Chandrawal Dam is situated on Chandrawal River at Charkhari in Mahoba Dist. in UP. It provides Irrigation to CCA of 19038 Ha of Mahoba Dist. through 43 km. main canal and 32.28 km. distribution system.

Actual Irrigation (UP)

Ha

	Bariyarpur Weir	*Chandrawal*	*Urmil*
1994-95	84900	7140	3189
1995-96	56560	7285	3855
1996-97	73120	6833	4209
1997-98	71630	3950	1839
1998-99	76230	6852	5266

Betwa River Basin

The 590 km. long Betwa rises from Raisen in MP (near village Barkhares, S-W of Bhopal) in the Vindhya Plateau at an elevation of 576 m. It meets Yamuna in Hamirpur Dist in UP. It is a rain-fed river with very high discharge during monsoon and extremely low discharge during summers. The basin lies between the N latitudes 22°51' and 26°0' and E longitudes 77°10' and 80°20'. The basin includes parts of a number of districts of Bundelkhand like Sagar, Tikamgarh, Chhatarpur (all MP), Lalitpur, Jhansi, Jalaun and Hamirpur (all UP). The catchment area of the basin is 43895 sq. km. of which 68.84 per cent is in MP and 32.16 per cent in UP.

Tributaries: Ghurari, Gairao, Barwa, Garrukha, Dhasan (Tr-Ur, Sukhani, Saprar, Rohini), Lakheri, Chainich, Parwaha, Jamini (Tr-Sajnam, Shahzad), Bina, Bearma (Tr-Arjun), Jamini (Tr-Sajnam and Shahzad).

Dhasan River: A tributary of Betwa, this river touches Lalitpur district at the SE tip and flows about 38 km. before it re-enters the neighbouring Tikamgarh district.

There is a small stream named **Rohini**, a tributary of river Dhasan, which flows in the north-east direction across the Mahroni *tehsil* in the south-west corner of the district. Rohini has a dam built for irrigation.

Jamini River: An important tributary of Betwa, it enters Lalitpur dist. cutting through the forest near Madanpur village, and flows northward for 45 km. leaving the town of Mahroni on its right. It then takes north-easterly bend and after 6 km. further comes to form the boundary of the district for about 60 km. It comes very close to Betwa just before it finally leaves the district. Jamini has been dammed within Lalitpur district.

Some Existing Major Projects in Betwa Basin

	1	2	3	4
Project	Paricha	Dhukwan	Matatila	Rajghat
Year of Completion	1885	1909	1964	Ongoing
Height above foundation, m.	17	15	46	43.8
Gross capacity, MCM	91.41	106.44	827.69	2172
Purpose	I	I	I,H	I,H
Max. discharge cap. of spillways, cumecs	21510	18451	23360	38997
Benefit, '000 Ha	—	1.31	5.46	109 (UP) 117(MP)
Catchment area, sq. km.		20824	20435	16861
Installed Power capacity, MW	—	—	30.6	45

I: Irrigation; H: Hydropower.

Source: For No. 1-3: Agro Economic, Socio Economic and Environmental Survey of Six Link Projects by NCAER and Large Dams in India, CBIP, New Delhi, 1987.

For No. 4: Storages in River Basins in India, CWC, New Delhi, 1987.

Other Important Streams: Important tributaries of Jamini are **Sajnam and Shahzad** rivers. While Sajnam joins it at Chandawali, Shahzad joins it near Hazaria village. These streams drain enormous volumes of water during rains while in other seasons they shrink to narrow channels. Shahzad, an important river flowing by the side of Lalitpur town, has been dammed to make Gobind Sagar reservoir near Lalitpur. Sajnam has also been dammed in this district.

Dams on the Tributaries of Betwa

Dam	*Completion Year*	*River*	*Ht. (m.)*	*Catchment (sq. km.)*	*Storage MCM*	*Dist. (State)*
Gobind Sagar	1953	Shahzad	18.29	368	96.80	Lalitpur (UP)
Shahzad	1992	Shahzad	18.00	514	130.0	Lalitpur
Sajnam	1990	Sajnam	18.78	290	83.50	Lalitpur
Jamini	1973	Jamini	19.18	414	92.89	Lalitpur
Rohini	1984	Rohini	15.50	44	12.12	Lalitpur
Barwasagar	1968	Barwa	—	—	10.2	Jhansi (UP)
Lachura	1910	Dhasan	17	NA	35.96	Mahoba (UP)
Pahari	1912	Dhasan	16	NA	79.34	Jhansi (UP)
Saprar (Kamla sagar)	1956	Saprar (trbty of SukhaniDhasan)	21	363.52	75.93	Jhansi (UP)
Siaori Lake	1911	Lakheri (Trbty of Dhasan)	14	—	7.82	Jhansi
Arjun	1957	Arjun	24	10.77	63.8	Mahoba (UP)
Keolari	1965	Keolari	11.73	N/A	7.67	Mahoba
Kabrai	1955	Magaria & Kulharni	15.25	N/A	13.22	Mahoba
Lakheri (U/c)	1988	Lakheri	10.6	N/A	15.6	Jhansi
Maudaha	N/A	Bearma	22	N/A	200	Hamirpur (UP)
Utari (Proposed)	U/C	Utari/Sajnam	N/A	80	11.16	Lalitpur (UP)
Bhaunrat (proposed)	U/C	Jamini	N/A	749.5	31.44	Lalitpur
Lachura Weir	1910	Dhasan	14.94	N/A	10.56	Mahoba (UP)
Majhgawan	1917	Gunchi Nala	17.07	N/A	26.8	Mahoba (UP)
Pahari Weir	1912	Dhasan	16.46	N/A	47.8	Jhansi (UP)

Paricha: Situated about 21 km. north-east of Jhansi on River Betwa. Paricha Weir provides Irrigation in Jhansi, Hamirpur and Jalaun districts (UP) through 30.2 km. main canal and 2625 km. distribution system. The CCA is 0.43 M Ha and annual irrigation is 0.23 M Ha.

Actual Irrigation

Ha

Year	*Paricha*	*Govind Sagar*
1994-95	279630	17453
1995-96	251510	17576
1996-97	238020	19272
1997-98	172210	8466
1998-99	217720	16688

Dhukwan: This weir on Betwa serves to enhance the irrigation capacity of Betwa canals.

Matatila: The dam is situated in Lalitpur district but does not provide any irrigation to Lalitpur district. The dam was constructed in 1957 for irrigation. Matatila power House with three machines of 10.2 MW each on the existing Dam has been commissioned in 1965. Matatila Dam provides 16.36 MCM water for drinking water supply. Benefited districts are Jhansi, Jalaun, Hamirpur (all UP) and Gwalior (MP). The total area submerged due to the project is 14243 Ha and total number of people displaced as per official estimates is 7500.

The rate of siltation in Matatila Dam has been three times higher than expected — increasing from 13200 to 44000 cum/100 sq. km./yr. The excessive siltation has decreased the reservoir capacity, reduced the life of the dam, and impeded the flow of water in the canals.

Due to construction of Rajghat dam, located about 50 km. upstream, Matatila dam would have to pass off an additional peak flood discharge of 38997 m^3/sec of water released from Rajghat.

Annual Generation

Year	*Generation in MU*
1996-97	131.172
1997-98	157.997
1998-99	132.805
1999-00	160.909
2000-01	138.083
2001-02	140.203

Rajghat: Rajghat Dam Project is an Inter-state project of the MP and UP being constructed on River Betwa about 22 km. from Lalitpur. The Engineering surveys were conducted in 1960 and the first project report submitted in 1970 and second report submitted in 1972, but the project could not materialise due to interstate problems. Finally an agreement was reached on 9 Dec. 1973 under which a tripartite Betwa River Board was establish under the then Union Minister of Irrigation with the chief ministers and concerned ministers of two states as its members. According to UP govt., the project on completion will provide irrigation to 109052 Ha in UP (districts Lalitpur, Jhansi, Jalaun and Hamirpur) and 116592 Ha in MP (districts Guna, Shivpuri, Datia Tikamgarh, Gwalior and Bhind). The water distribution system is under execution. The installed capacity of the powerhouse is 45 MW (3 x 15 MW). The costs and benefits of the project are to be shared by the two States equally. The projects existing on the downstream are Matatila Dam Project, Dhukwan and Paricha Weir. Thus, the Rajghat Dam Project will serve as mother storage for Irrigation in UP and MP through a cascade of hydraulic structures in the downstream of River Betwa. The total area submerged due to the project is 23390 Ha and the forestland submerged is 990 Ha. About 75 villages in UP and MP are partially and fully submerged, rendering 19000 people homeless as per official reports.

It is to create reservoir having a gross and live storage capacity of 2172 MCM and 1945 MCM respectively. If only 5 per cent of catchment area of the Rajghat Dam was considered critical, the cost of minimal Catchment Area Treatment would be over Rs. 48.1 million The cost of effective soil conservation measures would be more than Rs. 144 million, that is, 11.7 per cent of the Rs. 1.23 billion, which the project was expected to cost in 1973.

Apart from the cost of building the Rajghat reservoir (estimated at over Rs. 1.23 billion), the cost of canal irrigation system was additional.

There were at least 39 water bodies in the submergence area, ranging in size from 0.5 Ha to 19 Ha with an aggregate area of about 136 Ha. The Rajghat dam submerged about 23 temples and also Pancham Nagar Mahal.

Claimed additional irrigation for Bundelkhand Region includes 0.109 M ha in UP (districts Lalitpur, Jhansi, Jalaun and Hamirpur) —

Costing Rs. 2.25 billion and 0.117 M ha in MP (districts Guna, Shivpuri, Datia Tikamgarh, Gwalior and Bhind)— Costing 782.8 million. On completion of the project, the maintenance of Dam and regulation of reservoir shall be carried out by the Betwa River Board, whereas operation and maintenance of Rajghat HEP is to be done by the MPEB.

The estimated cost of the Rajghat Dam at March 1995 price level is Rs. 2.67 billion and at January 2000 price level is Rs 3.01 billion. The cost of the powerhouse at March 1997 price level is Rs. 1.31 billion The construction of the dam and its appurtenant works are almost complete. Land acquisition for 9 villages out of 49 villages in UP is to be completed. Civil, Electrical and Mechanical works of the Rajghat Power House have also been completed. All the three units of power House have been tested and synchronised during July to December 1999. According to Rajghat project agreement (1972), available water in river Betwa, at dam site was estimated 3303.8 MCM. As per studies conducted by UP in 1983, discrepancy was found in estimation of catchment area and also water availability at dam site was estimated as 3525.9 MCM. To meet the water shortage at Matatila dam, downstream of Rajghat dam, UP has requested to get the extra 222.1 MCM of water. Due to rectification in catchment area, MP had requested to Betwa River Board, for redistribution of water. Since the issue was beyond the jurisdiction of Board, it has referred to Central Regional Board.

Govind Sagar Dam: On Shahzad river, a tributary of Jamini river in Betwa basin, the GSD was completed in 1953 and has a catchment of 368 sq. km. It provides annual irrigation to 10830 ha through main canal of 66.43 km. and 123.57 km. distribution system. It provides 2.26 MCM water for drinking water supply. Live storage is 80 MCM.

Shahzad Dam: In Lalitpur district, just upstream of Lalitpur town, on Shahzad river, a tributary of Jamini river in Betwa basin. Shahzad dam, completed in 1992, provides irrigation to 16002 ha in Dist Lalitpur through main canal of 43.75 km. and distribution system 56.1 km. It provides 1.7 MCM water for drinking purpose. Catchment area is of 514 sq. km., live storage 96 MCM.

Sajnam Dam: Sajnam dam provides annual irrigation to 7145 ha in Dist. Lalitpur through main canal of 37.5 km. and distribution system 48.05 km. It provides 1.42 MCM water for drinking purpose.

Completed in 1990, it has live storage of 75 MCM. It has catchment area of 290 sq. km. on a tributary of Jamini River.

Actual Irrigation (UP)

Ha

Year	*Sajnam*	*Jamini*	*Shazad*	*Lahchura Weir*
1994-95	12951	32710	9455	47856
1995-96	13136	32757	10043	41186
1996-97	13162	33333	10223	40821
1997-98	3158	3150	6635	31978
1998-99	13656	32786	8457	35971

Jamini Dam: Jamini Dam, completed in 1973, provides irrigation to CCA of 55144 ha through main canal of 67.4 km. and 177.91 km. distribution system. It provides 1.42 MCM water for drinking water supply. Annual Irrigation is of 13699 Ha. Catchment area is 414 sq. km. Live storage is 84 MCM. Jamini River is a tributary of Betwa River.

Rohini Dam: Situated in Lalitpur dist., it was completed in 1984. Rohini Dam provides irrigation to CCA of 3302 ha in Dist Lalitpur through main canal of 8.64 km. and distribution system 11.22 km. It provides 0.71 MCM water for drinking water supply. It has catchment area of 44 sq. km., gross storage of 12.12 MCM. Rohini is a tributary of Dhasan River.

Barwasagar: Barwasagar is situated nearly 15 km. east of Jhansi and dammed by a 1.21 km. strong embankment, this lake dates to the Chandela period. It was rebuilt by Orchha ruler in the 18th century to hold 10.332 MCM water. Canals from this lake were built sometime before 1862 AD.

Siaori Lake: Situated at about 8 km. north-west of Mauranipur at village Siaori on Lakheri river, this lake was improved in 1906 and opened for irrigation. This also receives water from Kamlasagar, which has increased its irrigation capacity.

Lahchura Weir Lachura Weir in Dist. Mahoba is built on Dhasan at about 11 km. north of Harpalpur station during 1906-10. This dam consists of a masonary weir and earthen bunds on each side. It has CCA of 97790 ha in Mahoba and Hamirpur districts through main canal of 32 km. and distribution system of 599.63 km. The live capacity of this 14.94 m. dam is 10.56 MCM.

Saprar Dam: The Reservoir is known as Kamlasagar. Built during the First Five-Year Plan, this consists of a 3.9 km. long earthen dam. The reservoir irrigates Jhansi district through Ranipur canals and enhances the capacity of Siaori lake.

Pahari Dam: Situated about 18 km. east of Mauranipur in Jhansi district on Dhasan river this weir was built during 1909-12. This serves the purpose of irrigation through the Lachura Dam mainly in Hamirpur district. 16.46 m. Pahari Weir provides irrigation to Jhansi dist. Gross capacity of reservoir is 47.8 MCM and live storage capacity is 46 MCM.

Arjun Dam: Arjun Dam is situated at Charkhari in Dist. Mahoba on Arjun River, which was completed in 1957. Arjun Dam provides irrigation to CCA of 59722 ha of Dist. Mahoba and Hamirpur through main canal of 42 km. and distribution system 217.3 km. Length and height of the dam are 5200 m. and 27.43 m. The gross and live capacity of reservoir is 68.35 and 62.97 MCM.

Actual Irrigation

Ha

	Arjun Dam	*Majhgawan*	*Keolari*	*Kabrai*
1994-95	18264	7505	3796	3684
1995-96	18778	7297	4113	3663
1996-97	19501	7551	4044	4104
1997-98	8225	3667	761	2831
1998-99	23244	7513	1568	4520

Maudaha Dam: The 22 m. high Maudaha Dam on Bearma River, a tributary of Betwa in Rath Tehsil in Hamirpur district. It consists of 48 Km. long main canal with head discharge of 15 cumecs, 50 Km. of remodelled Channi and Sumerpur branch canals with distribution network of 337.8 Km. It is to provide irrigation to 24297 Ha in *Rabi* and 3937 Ha in *Kharif* in Hamirpur and Mahoba districts. It provides 2.8 MCM for drinking water supply. The 200 MCM gross capacity reservoir has live storage of 179 MCM.

Majhgawan Dam: Majhgawan Dam is situated in Dist. Mahoba on Gunchi Nala, a tributary of Dhasan and the project was completed on 1917. Maximum flood discharge from the dam is 170 cumecs. Majhgawan Dam provides irrigation to 11248 ha. in Dis.t Mahoba and Chhatarpur (MP) through main canal of 29.1 km. and distribution system 31.30 km. Height of the dam is 17.07 m. Gross capacity of the reservoir is 26.8 MCM and live storage capacity is 26 MCM. Irrigable Command Area is 11248 ha.

Keolari Dam: On Keolari River, a tributary of Dhasan in Mahoba district of UP, Keolari Dam provides irrigation to CCA of 14390 ha and annual irrigation of 4100 Ha. through main canal of 28.27 km. Completed in 1965, the reservoir has gross storage capacity of 7.67 MCM and live capacity of 7.23 MCM.

Kabrai Dam: Completed in 1955, it provides Irrigation to CCA of 14960 Ha. and annual irrigation of 3760 Ha. in Dist. Mahoba and Hamirpur through main canal of 23.544 km. and distribution system 66.4 km. It provides 1.73 MCM water for drinking water supply. The Gross storage is 13.22 MCM and live storage 11.94 MCM.

Some under-Construction Projects — Lakheri Dam: The Dam is situated a little upstream of the junction of Chiraya and Tola Nallas near Village Mahewa about 16 km. from Mauranipur in Dist. Jhansi on Lakheri river. Maximum flood discharge of the dam is 1744.07 cumecs. The construction of the dam started in 1981. Lakheri Dam will provide irrigation to 1980 ha of land in Doab of Lakheri and Pathari river spread in 13 villages of *Tehsil* Garautha, through main canal of 9.20 km. and distribution system 21 km. The length and the height of the dam are 4880 m. and 10.6 m. respectively. Dead Storage Capacity of the dam will be 1.7 MCM and Live Storage Capacity will be 13.9 MCM.

Proposed Projects — The Orchha Multipurpose Project: This Project was proposed by MP on river Betwa in Tikamgarh district in 1978. The project costing Rs. 668.2 million is to produce 90 MW hydropower and irrigate 29150 ha. On the same river Dhukawa HEP was proposed by UP, utilising the water head between Dhukawa and Paricha, with an installed capacity of 75 MW.

Proposed project of MP will require construction of three dams and two barrages, whereas Dhukawa project of UP will require neither construction of dam, nor any submerged area. In view of MP, Orchha Multipurpose Project is more beneficial, as it would generate excess of 15 MW power and would irrigate 29150 ha. area. In secretary level meeting held at Bhopal in April 1999 it was decided to undertake comparative studies of both the projects. In compliance of decision taken in Central Regional Board meeting in February 2000, a questionnaire for comparative studies of the projects has been sent to MP and the response is awaited.

Other Proposed Projects: The centre and state governmentts have sanctioned Rs. 232.5 million for power generation from various small HEPs in UP. Betwa Project is expected to produce 0.2 MW and the Betwa Canal Head powerhouse in Jhansi is expected to generate 1.3 MW. The Ghunchai Project in Pilibhit would produce 0.8 MW, while 0.2 MW would be generated from Jamini project in Lalitpur and 1.4 MW from Kuthaund Project (Jalaun). (BUSINESS LINE 19-04-03)

- CEA has done ranking study for in India's Hydropower schemes. Two of Project named Orchha (39 MW, UP) and Dhurwara (28 MW, MP) are also proposed in Betwa basin.

Utari Dam: It is proposed on river Utari, a distributary of River Sajnam near village Surikalan in Mahroni Tehsil of Dist. Lalitpur. Utari Dam will provide irrigation to CCA of 2012 ha. through feeder channel of 10.8 km. length from its right flank. It is to provide *kharif* irrigation of 600 Ha. and *Rabi* irrigation of 1800 Ha. The Gross storage capacity is 11.16 MCM and live capacity 10.82 MCM.

Bhaunrat Dam: It is proposed on river Jamini 20 km. D/s of existing Jamini Dam near village Bhairoghat in Mahroni Tehsil of Dist. Lalitpur. Bhaunrat Dam will provide irrigation to CCA of 7900 Ha through feeder channel of 17.6 km. from left bank. It has gross storage capacity of 31.44 MCM and live capacity of 29.75 MCM. It proposes to provide *kharif* irrigation in 2500 Ha. and *Rabi* irrigation in 7900 Ha.

The Link Proposal

The Ken drains an area of 28060 sq. km. of which 16020 sq. km. is cultivable. The water balance carried out by NWDA indicates that the Ken basin as a whole has a potential of 10968 MCM of surface water resources. Based on the Technical Advisory Committee 60 per cent of the cultivable area (9870 sq. km.) will be brought under irrigation by 2025 AD. Thus, the requirement of surface water will be 5883 MCM in the entire basin. Thus, it is claimed surplus of 5085 MCM in the Ken basin.

Salient Features of the Proposed Project

Location: Latitude 24°37' 30" N, Longitude 79°51' 40" E

Greater Gangau Dam

MWL	288 m
FRL	287 m
Minimum Draw Down Level	268 m
Dead Storage Level	238 m
River Bed Level	215.28 m
Water spread at FRL	8650 Ha
Water spread at MWL	9000 Ha
Live storage	2752.69 MCM
Capacity at FRL	2775 MCM
Masonary Dam: Foundation Level	209.94 m
Top of Dam	291 m
Earthen Dam: Top of dam	287.97 m
Max ht. above Ground	24.49 m
Link Canal: Bed width at head	12.00 m
Full depth of flow	3.50 m
Side slope	1.5: 1
Length	231.45 km
Land required for Canal in UP	2135 ha in MP and 180 ha
Design discharge	72.00 cumecs upto 134 km, 62 cumecs 134 - 195 km & 57 cumecs 195 km-terminal at BarwaSagar
Affected Area and Persons (only by GGD)	
No. of Villages affected by reservoir	19
No. of persons affected	8550
Forest area	6400 Ha

Dams in Raisen and Vidisha Districts (MP)

Proposed Projects		Sq. Km to be Irrigated as Per Master Plan	Annual Irrigation (sq. km.)*	Water Requirement (MCM)
A. Betwa Complex	Barari Barrage	696	870.09	452
	Neemkhera Dam	8	10.53	5
	Richan Dam	295	368.28	192
B. Kesari Dam		21	18.40	10
Total upper Betwa (A+B)		1020	1267.30	659

* As estimated by NWDA.

In addition there is also a 60 MCM dam on Kainu Nadi, a tributary of Betwa river. Thus, in all there are going to be six large dams to be created under the KBLP, but even basic information about any of the dams except the Greater Gangau Dam has not been given.

Sub-Basin-wise Surface Water Resources of Betwa Basin (sq. km.)

Area and Water Resources	Betwa Basin		Jamini Basin	Dhasan Basin	Bearma Basin	Total
	Lower	Upper				
A. Geographical Area	8635	16876	4510	11102	2772	43895
B. Culturable Area (% of A)	6266 (72.57)	10977 (65.05)	3251 (72.08)	6909 (62.23)	2391 (86.26)	29794 (67.88)
C. Claimed Gross Irrigated Area by 2025	2424 (38.68)	6089 (55.47)	1607 (49.43)	2450 (35.46)	912 (38.14)	13482 (45.25)
D. Total Availability as Surface Water Resources, MCM (D/A)	2497.0 (0.289)	4676.0 (0.277)	1156.0 (0.256)	2912.1 (0.262)	717.7 (0.259)	11958.8 (0.272)
E. Requirement of Surface Water Resources, MCM (E/C)	2855.0 (1.18)	6172.0 (1.01)	1101.0 (0.685)	2762.8 (1.13)	829.5 (0.91)	13720.4 (1.018)
F. Surplus/Deficit of Surface Water	-358.0	-1496.0	+55.0	+149.3	-111.8	-1761.6

Sub-Basin-wise Surface Water Resources of Ken Basin (sq. km.)

SN	*Details*	*Ken Basin*		*Vearma Basin*	*Sonar Basin*	*Total Basin*
		Lower	*Upper*			
A.	Geographical Area	8722	6986	5890	6550	28058
B.	Culturable Area (B as % of A)	6735 (77.22)	3232 (46.26)	2753 (47.74)	3295 (50.31)	16015 (57.08)
C.	Gross Irrigated Area by 2025 AD (C as % of B)	5890 (87.45)	1387 (42.91)	1020 (37.05)	1576 (47.83)	9874 (61.65)
D.	Total Availability of Surface Water, MCM (E/A)	4453.3 (0.5106)	2165.0 (0.3099)	2480.8 (0.4212)	1868.8 (0.2853)	10967.9 (0.391)
E.	Requirement of Surface Water, MCM, (E/C)	3412.9 (0.579)	782.3 (0.564)	736.4 (0.722)	952.1 (0.604)	5882.8 (0.596)
F.	Surplus/Deficit of Surface Water	(+)1040.4	(+)1382.7	(+)1744.4	(+)916.7	(+)5085.1

Ken-Betwa Surface Water Balance

(MCM)

	Ken	*Betwa*	*Total*
1. A. Total SW available at 75% dependability	7657.6	9196.2	16853.9
B. Import	2426.9	955.2	3382.1
C. Regeneration	883.4	1807.4	2690.8
TOTAL	10967.9	11958.8	22926.7
2. Requirement SW			
A. Domestic Use	220.4	1103.3	1323.7
B. Industrial Uses	402.8	903.5	1306.3
C. Irrigation	5260.3	8301.0	13561.3
D. Exports		3854.5	3854.5
TOTAL	5882.8	13720.4	19603.2
3. Surplus (+)/Deficit (-)	(+)5085.1	(-)1761.6	(+)3323.5

Surface Water Balance Upto Greater Gangau Dam Site

(MCM)

A. Gross annual yield at 75% dependability		6211
B. Surface water requirement for	Irrigation	2969
	Domestic	133
	Industrial	236
	Sub Total	3338
C. Export (for downstream Irrigation needs)		2225
D. Regeneration from	Irrigation	176
	Domestic	107
	Industrial	188
	Sub Total	471
E. Net Availability (A-B-C+D)		1119

It has been proposed to construct four large reservoirs, namely, Barari Barrage and Neemkheda Dam on the main Betwa, Richhan dam on Richhan river and Kestan Dam on Kestan river. These reservoirs will have an annual irrigation potential of 0.127 M. Ha. in Raisen and Vidisha districts of MP. The water balance conducted by the NWDA is based on 1901 to 1983-84 data.

The proposed link canal will take off on the left bank of the Ken river at 245.5 m. from the tail race of power house — I of the GGD. The GGD site is covered by hillocks with dense forest. The proposed link canal will run in almost northerly direction for distance of 81 km. in MP thereby passing through the Bhusor and Bandari protected forest area of Chhatarpur district. Thereafter it will run in westerly direction for a distance of 131.4 km. within the vicinity of the state boundary between MP and UP. The Canal will cross on its way the Dhasan river, a major tributary of Betwa, many other small minor streams, state highways, railway line and Pabra and Magarwara reserve forest till it outfalls into the terminal reservoir across the Kainau Nadi near village Jobra (as per NCAER Report. However, the NWDA Feasibility Report does not mention this). This terminal reservoir will have a storage capacity of 60 MCM. The sluice in the terminal reservoir will regulate the flow of the water to Barwasagar Nalla, which will connect to the Betwa river at a point 13 km. upstream of the existing Paricha Weir. As per the agreement between the governmentts of UP and MP, these states will receive 850 and 1375 MCM of surface water respectively from the GGD. (This possibly refers to existing irrigation from Ken River downstream of the existing Gangau Dam, as shown above in the surface water balance downstream of existing GD).

Proposed Command Area of Link Canal

The Ken-Betwa link canal project proposes to provide enroute irrigation in 89 villages in the Chhatarpur and Nowgang *tehsil* of Chhatarpur and 74 villages in Niwari and Jalara *Tehsil* of Tikamgarh districts of MP. It is proposed to provide irrigation to the areas/ *tehsils*, which have less than 30 per cent annual irrigation of their culturable area by 2025 AD.

Extension of Irrigation in the Upper Betwa Basin

Based on the agreement between the MP and UP on sharing of Betwa waters, the MP government has prepared a master plan to utilise 659 MCM of surplus water by constructing the Barari Barrage, and the Neemkhera, Richhan and Kesari Dams in the upper reaches of the Betwa to provide irrigation facilities in the districts of Raisen and Vidisha in MP. To replenish the water used in the upper reaches of the Betwa, an equal quantity of water (659 MCM) will be made available through the proposed Ken-Betwa link canal. It is proposed that the upper Betwa complex will provide for the irrigation of 0.127 M Ha a year, given the 125 per cent intensity of irrigation. That means 5189 cubic m. of water is provided per Ha irrigated.

On account of the transfer of surplus water of the Ken to Betwa 13 km. upstream of the Paricha Weir, the *tehsils* now having less than 30 per cent culturable area will by 2025 AD derive the benefit of irrigation (Mahoba, Muranipur, Jhansi, Konch and Hamirpur). The annual irrigation of 69194 Ha and 365 MCM of water utilisation will take place in these *tehsils*. That means 5275 cubic m. of water is provided per Ha irrigated.

Cost Estimates

The entire Ken-Betwa link project has been divided into two units. Unit-I relates to the Ken-Betwa link project works and unit-II to the link canal and the other canal networks, which also includes the upper reaches of the Betwa and the lower Betwa basin down stream of the Paricha Weir up to the confluence with the Yamuna.

Unit III includes two power houses, power intakes structures, tunnels and penstocks, surge tank, tail race and approach tunnel, switch years and transformer yard, transmission line, cost of electric equipments, etc.

Cost Estimates Ken-Betwa Link Canal

(Rs. billion)

	1989-90 Price Level	*April 1994 Price Level*
Unit-I	1.59	3.6792
Unit-II	0.3995	5.7244
Unit-III	—	0.5072
Ken Command	—	5.5411
Betwa Command	—	4.4356
Total	1.99	19.8875

Note: 1989-90 price level figures were given in the NCAER report and the 1994 price level figure are given in NWDA's Feasibility Report. Betwa command cost figures are notional, calculated on the basis of assumption that it will cost Rs. 35000/- per ha of area irrigated.

Currently (September 2005) the estimated cost has gone up to Rs. 45 billion as per Union Minister for Water Resources.

Environmental Implications

Out of six dams to be constructed under KBLP, information is given in the NCAER study about only one of the dams, namely GGD. No information is given about the other dams. So the total impacts of all the proposed dams and canals under the KBLP cannot be known till full information about all the dams and canals is given.

The left bank of the site selected for GGD is gradually rising hill slope and the right bank is steeply rising hill slope. Number of streams flow down the hill slopes forming the gulleys. The estimated areas to be submerged due to the formation of reservoir at the designed full reservoir level is 9605 Ha (p 16, Vol. II) or, almost 10 000 Ha (p 17, Vol. II) coming under Panna, Chhatarpur and Damoh dists. of MP. Channels near the reservoir are likely to get enlarged and loosen the talus and screen. This may cause increase in silting. However, the submergence area, strangely comes down to 8650 ha in the Feasiblity Report, even though the height of the dam goes up in the Feasibility Report. No explanation is given for this difference, which gives rise to a suspicion that the submergence area has been understated.

With FRL at 284.2 m., the submergence area comprises of 3750 Ha forest, 2510 Ha cultivable land and 3740 Ha others. Part of the area, close to the reservoir and to the south along Ken river, will fall under Panna district and the western part along Shyamri river will fall under Chattarpur and Damoh districts in MP. As per the toposheet study the villages which will get fully submerged including the habitations are Daudhan (dam site), Kharyani (5 km. south of Daudhan), Palkoha (4.5 km. south west of Daudhan), Sukwaha (6 km. SW of Palkoha), Bhorkhuwa (3.5 km. SW of Sukwaha), Basudha (5.5 km. SW of Bhorkhuwa) and Ghughari. About 30 km. length of road (Gangau-Palkoha-Sukwaha-Bhorkhuwa-Basudha-Shahpura) will also get submerged. Maniyari and Padriya villages will also get affected, but NCAER claims they are part of the above-mentioned villages. Many other villages will be affected, as admitted by NCAER, but NCAER has no information about them. At one place (page 77, Vol. II), NCAER mentions that 19 villages will be affected, but no details are given about affected population.

The protected forest slated for submergence on the north is away from the reservoir area but a part in south is within the protected forest included in the Panna National Park. Almost all the hill slopes are marked by dense mixed jungle comprising deciduous, a few evergreen to semi-evergreen and few xerophytes to semi-xerophytes type of species.

It is claimed that the canal will be designed with proper cross drainage works. The movement of subsurface groundwater may get affected along the route of canal. It is claimed that the link canal will be fully lined with suitable material.

The villages along the periphery of the reservoir will face the problem of wastewater disposal, as the existing drains will become inefficient due to the reservoir in their vicinity. The estimated number of families in the villages to be submerged is 600 in NCAER report and 900 in FR. The population, thus, facing submergence is given as 3250 in the NCAER report and 8550 in the FR. Scheduled tribes and Scheduled castes constitutes about 34.38 per cent and 15.54 per cent respectively, as per NCAER report. These are the villages in forest areas. In the interior villages the concentration of scheduled tribes is very high e.g. 91.84 per cent in Ghughari village. Population in the other peripheral villages will also be affected but no details available in this regard.

At present the groundwater is used by villagers from the wells in the vicinity of the villages and the springs nearby. The wells are generally shallow (3-6 m) and subterranean water is available in them.

The total peripheral zone around the area likely to be submerged in the reservoir could not be examined by NCAER during its study of the Socio Economic and Environmental impacts of the project, due to their inaccessibility. However, only partial study of the area was done and geological and hydro geological data for the rest is used for assessing the impact on the ground water regime. The function of rivers as good surface drainage outlets for these villages will be affected adversely. The groundwater table will rise and general deterioration of the environment will take place.

In the command area water-logging may result unless proper water distribution plan is prepared and implemented. Certain areas on the existing roadside over which the canal will pass, are dependent on well water and shallow tubewells located on crushed and altered granitic and gneissic rocks. The catchment areas of such rocks are likely to get covered by the canal construction, resulting in reduction in drinking water supply.

The Manipulated Water Balance

The water-balance that is at the basis of the KBLP as described above is fundamentally flawed in many respects, some of which are described below:

- The most fundamental problem with the water balance study is that it does not take into account groundwater potential and use in the relevant basins. Moreover, the water balance also does not look at rainwater as a resource and potential of rainwater harvesting before deciding if there is really any deficit or surplus.
- Another very fundamental issue is that there seems to be no allocation for the environment flow requirements in the river downstream from the six proposed dams.
- While calculating surplus in Ken basin, the assumption is that every Ha irrigated will need 5327 cubic m. water. This is very low compared to the requirement assumed at 6157

cubic m. per Ha assumed in case of Betwa basin. No reason is given for this huge 16 per cent difference. However, if we look at the fact that whole attempt is to show that Ken is water surplus basin and Betwa is water deficit basin, the rationale behind such assumptions becomes clear. By assuming low water requirement in Ken basin, one can show surplus water availability and by showing higher water requirement in Betwa basin, one can show higher deficit in Betwa basin and thus try to justify a link project that otherwise has no justification.

- This attempt to show the surplus in Ken basin and deficit in Betwa basin is further exemplified by the fact that 67.88 per cent of geographical area in Betwa is shown to be cultivable, the figure for Ken basin is much lower at 57.08 per cent.
- About 85 per cent of the so-called water deficit in the Betwa basin is seen in the Upper Betwa basin, where water from Ken Betwa link cannot be taken. This deficit has been made possible by the assumption that 65.05 per cent of Upper Betwa sub-basin is cultivable and 55.47 per cent of cultivable area here is to be irrigated by 2025. The percentage cultivable area to be irrigated is highest for Upper Betwa sub-basin among all the sub basins in Betwa basin. Contrast these with the figures for Upper Ken Basin: 46.26 per cent of geographical area is cultivable and 42.91 per cent of cultivable area is to be irrigated by 2025.
- It is assumed that by 2025, 87.45 per cent of cultivable area of Lower Ken sub basin will be irrigated, while only 37.05 per cent of cultivable land in Vearma basin will be irrigated. This clearly shows that the projections are to provide more irrigation to already highly irrigated areas, and starve the unirrigated areas. This difference becomes even more alarming when we consider the fact that over 77 per cent of geographical area in Lower Ken basin is considered cultivable, whereas the figure for Vearma basin is just around 47 per cent.
- The Study mentions in the Betwa basin water balance that some 3854.5 MCM is to be exported from the Betwa basin. No information is given what is this about, from where and

to where and what for this export is. The moot point is that if Betwa basin were not to export this amount of water, it would in fact be a surplus basin. Similarly, no details are given about 2426.9 MCM water being imported into Ken basin and 955.2 MCM water being imported into Betwa basin, as to from where, how, where exactly in the respective basins and what for these imports are. It is clear that NWDA and NCAER has refrained from giving correct data and has in fact tended to give biased information, giving rise to a suspicion that the figures are manipulated to justify an otherwise unjustifiable project.

- This bias is further evident when the study repeatedly describes Betwa as "water starved" basin right at the outset (e.g. Vol. II, page iii, again on Vol. II page 6), whereas the role of (NWDA and) NCAER is to study and investigate what is the situation of water resources in both the basins.
- The ignorance of NCAER and shoddiness of the report is apparent when it states in Executive Summary (Vol. II page iii again Vol. II p 3) that KBLP will benefit the districts of Ujjain and Indore in MP!
- The shoddy piece of work that NCAER has done is further evident when on page 3 (Vol. II) it says that KBLP involves "construction of large reservoirs on river Yamuna."

Questions on Projected Irrigation Benefits

The irrigation benefits projected from the proposed KBLP seem highly doubtful. Some reasons are given above when there are fundamental discrepancies in the water balance studies. The fact that utilisation of rainwater and groundwater in the Ken-Betwa basin is so low shows that there are better options available if fulfilment of justifiable needs of irrigation water in Ken-Betwa river basins are the objective.

Moreover, if we look at the water allocation and area to be irrigated in KBLP, we find that water allocation is 5189 to 5275 cubic m. for every Ha to be irrigated under KBLP. It is a mystery that if the Betwa basin water balance assumes irrigation requirement of 6157 cubic m per Ha of irrigation provided, than why is just about 20 per cent less water allocated for area irrigated by link canals in the same

Betwa basin? Or is this just to push up the projected irrigation areas to levels beyond what can be achieved? This is another indication of how the FR is essentially an attempt to push unjustifiable project?

What studies do not say

The impacts that the submergence will have on the Panna Tiger Reserve (PTR) will be numerous and serious.

According to official website of the PTR, the Ken river, which flows through the Reserve from south to north, is home for Gharial and Mugger, and other aquatic fauna and is one of the least polluted rivers. It is one of the sixteen perennial rivers of Madhya Pradesh and is truly the lifeline of the Reserve. Ken offers some of the most spectacular scenery to the visitor while it meanders for some 55 km. through the reserve. *(http://www.pannatigerreserve.org/)* When Great Gangau Dam comes up, the reservoir will not only submerge significant parts of PTR, it will also make approach of the wildlife to the only perennial water source impossible due to the silt that gets deposited on the periphery of the reservoir. Moreover, the Ken Gharial Sanctuary, located downstream of the proposed KBLP is also likely to be affected due to stoppage of freshwater flow in the river.

Also, destruction of over 3750 Ha of rich forests due to GGD alone will have many impacts by itself. This will mean loss of wildlife and all the biodiversity, loss of whole system of rainwater absorbing ecology and subsequent increase in water flow in the monsoon and decrease in the water availability in non-monsoon months, shrinkage of resource base of surrounding population and shrinkage of living space for the wildlife.

Says the official website of MP tourism dept *(http://www.mptourism.com/dest/khaj_exc.html)* about PTR, "Most likely, it will be here if the Caracal, a vanishing cousin of the extinct Cheetah, is ever sighted." The building of GGD will most likely forever submerge such a prospect.

According to official website *http://panna.nic.in/tiger.htm,* PTR has the following Endangered Species:- Tiger, Leopard, Caracal, Four-horned antelope, Indian Wolf, Pangolin, Rusty Spotted Cat, Sloth Bear and Gharial, found in the Park, are included in Schedule I of WPA, 1972. Their habitat will be seriously impacted by the proposed GGD.

Does NWDA or NCAER know the meaning of EIA?

In the beginning of the chapter 2 of Vol. II, NCAER has listed what it calls "massive data" required for "Comprehensive assessment of environmental impact." That list *excludes*, believe it or not, command area impacts (water-logging, salinisation, drainage), downstream impacts, impacts on biodiversity, loss of forests, carrying capacity, impact of sudden releases in the downstream areas, geologic, seismic issues, siltation and catchment area treatment, to name just a few. This strongly raises the question if the NCAER understands the meaning of EIA and it is capable of taking up such studies. The ILR task force, we understand, has asked NCAER to take up the EIA work! That would certainly seem a disaster in waiting.

R&R for Link Projects

According to NCAER study, NWDA has not prepared any concrete proposal for the rehabilitation and resettlement of the population likely to be affected. Therefore, the persons likely to be affected are not aware of any R&R package.

On KBLP, the NCAER study says the total number of families to be displaced will be 600 and number of persons would be 3250. In Vol. I of the NCAER study it is shown that in this area, there are on average 6.5 persons per family. This means that 3900 persons will be displaced if 600 families are to be displaced. However, if population densities in the area is any guide, than at least 10000 families (65 000 population) will be displaced by one GGD alone. No information is available about submergence due to five other dams or the link canal area or in the downstream areas or due to building of other related infrastructure.

While in Annexure 7 of Vol. II NCAER tries to give the R&R norms followed in various states and various projects, it either does not know the provisions or it is trying to misguide the readers. In either case, NCAER becomes disqualified for any such work in future. To give just one stark example, NCAER says that in case of Sardar Sarovar Project, norm is to give land for land upto 4 acre land holding, 4 acre land for 4 to 12 acre holding and 1/3rd of the holding for 12-15 acres land holding and so on. But the norm is that all families, including the encroachers and the landless in the submergence villages are to get a minimum of 5 acre of irrigated land (irrigation to be provided by state government) and each major son (or daughter in some states) is considered a separate family. Even landless are to get that provision of land. Thus, by not stating the existing provisions honestly, NCAER has shown its bias against the affected people.

What is clear is that total quantum of submergence, displacement is not even known to either prepare any just R&R plan or take the affected people into confidence and take their free, prior and informed consent about the project or any of the options for irrigation needs of the region. Also, the people displaced by the dams in the Ken and Betwa basins described in earlier sections of this report are yet to be fully and justly resettled. Unless that is achieved, any further displacement in the basin cannot be considered, as recommended by the report of the World Commission on Dams.

Public Awareness

An attempt has been made by NCAER to find out awareness among the households regarding the proposed link project in the command area. Only 8 per cent were aware regarding the proposed project while 92 per cent were not aware. This clearly shows that even as pre-feasibility and Feasibility Reports of the scheme are claimed to have been done, local people are not even informed about the existence of the proposal.

Perception about Irrigation

During the course of field inquiry an attempt was made by NCAER to ascertain the extent of existing irrigation facilities along with various socio-economic and agro-economic parameters in the command areas of the proposed link project. According to the qualitative view of irrigation facilities in the command area 72.4 per cent households stated that the existing irrigation facility is not adequate, while 27.6 per cent households stated that existing facility is adequate. According to NCAER, this suggests urgent need of the creation of additional irrigation potential. This clearly is a wrong and biased conclusion and also a misrepresentation of the results. The question was about quality and not only quantity of irrigation. Moreover, if irrigation facilities are quantitatively inadequate, there can be very many ways of extending it and link canal is not the only or even the best way.

The households of command area of link canal were asked to state their suggestions for improving irrigation facilities in their areas. The responses are as follows:

SN	*Irrigation Facility Required*	*Per cent*
1	Canal	11.8
2	Lift Irrigation	49.2
3	Well/Tubewell	26.8
4	Ponds	5.3
5	Others	6.5
6	Not responded	0.4
	Total	100

The majority of respondents suggested for lift irrigation and well and tubewell as a solution of inadequate irrigation facility. This clearly shows that majority of the people were not for a project like KBLP. In fact only 11.8 per cent households have preferred canal irrigation as additional irrigation facility. The vote of the local people is clearly against KBLP.

Possible Impacts in Ken Basin

The proposed link canal would submerge thousands of hectare cultivable land at Chhatarpur, Tikamgarh and Jhansi. Gangau and Bariyarpur dams are situated at downstream of proposed dam. The irrigated area from both projects could get seriously affected due to the proposed link. Thousands of Ha of cultivable land would be taken for canals in Chhatarpur and Tikamgarh. People from scores of villages near Dhaudhan would be displaced due to proposed dam. A major part of world-fame Panna Tiger Reserve would also be submerged due to the reservoir. The Gangau Dam is the feeder dam for Bariyarpur Weir, which will receive less or no water from the Ken River during non-monsoon months. Even in monsoon, while the Greater Gangau Dam would be filling, there will be no water either for Gangau Dam or for the Bariyarpur Weir and its command areas. Hence Gangau and Bariyarpur reservoir may become dry for most of the year if all the water in the Ken River is stored in the GGD and is diverted to Betwa basin.

Impact in Betwa Basin

The excess water in Betwa basin could create water logging in Hamirpur, Mahoba, Banda and Jalaun districts. The excess water could also make Hamirpur (Including Mahoba), Banda and Jalaun districts flood prone. The groundwater development in all three districts is very low, at 10 per cent in Banda and 12 per cent in Hamirpur and Jalaun districts.

- These districts are situated downstream of Barwasagar reservoir. The combined Hamirpur district has an area of 7165 sq. km. Area affected by soil erosion in both districts is 364218 Ha, which is 50.56 per cent of the total area.
- The total area of Banda district is 7624 sq. km. There are seven major watersheds in this district, each has large eroded area. The total area affected by soil-erosion in the district is 1.2 M. Ha.
- The total area of Jalaun district is 4565 sq. km. and the total cultivated area is 341 818 Ha. The area irrigated by minor irrigation works is 114 000 M. Ha.

UP – MP Disagreements

Fundamental differences have persisted between Uttar Pradesh and Madhya Pradesh Government on the proposed KBLP and its earlier incarnations of Ken Multi Purpose Project. It is a mystery how UP government agreed to sign the MoU on August 25, 2005, in spite of these serious differences. It was said at the time of signing of the differences will be taken care of during preparation of the DPR. However, how can such fundamental differences that raise questions about the viability of the project be resolved? And when such questions exist, why spend huge amounts like Rs. 40 crore for taking up the DPR.

At a meeting on January 11, 2005 between the Chief Secretaries of Government of UP, MP and Rajasthan under the chairmanship of the Secretary, Union Ministry of Water Resources, the objections raised by UP Chief Secretary and the UP Principle Secretary for Irrigation to the KBLP included the following. It is clear that the objections being raised by UP, according to official minutes of the meeting, are very serious in nature and the KBLP would have far reaching impacts in Ken and Betwa river basins. It is also equally interesting to see how Secretary, MWR, Government of India is responding to the issues raised by UP.

1. **Ken is not a surplus basin:** Principle Secretary, Government of UP made it clear that Ken Basin is not surplus basin and if water is transferred from this basin there might be unrest in the Bundelkhand region. This is indeed the most fundamental issue. Secretary, Union MWR's response is amazing: This issue has already been discussed and 'need not be opened again.' NWDA response at the meeting is equally strange: As per NWDA criteria, Ken is a surplus basin. So an issue that raises the most fundamental objection to the Ken Betwa link proposal by the Principle Secretary, UP, has been pushed aside by the bureaucrats without even attempting to answer them. As we have shown in this critique, indeed NWDA has manipulated the water balance figures to show that Ken is a surplus and Betwa is a deficit basin, whereas in reality there is little difference between the situations in two basins. The warning of social unrest in Bundelkhand region has not even been addressed. This is another piece of evidence that shows that the government is trying to push a project that has fundamentally no merit. What is also shocking here is that UP government officials, in stead of demanding that this issue be resolved first, agreed to go ahead with further discussions and latter on to sign the MOU, keeping the people of UP entirely in dark.

2. **Irrigation in Lalitpur and Jhansi to be affected:** UP officials said that the areas presently irrigated in UP, south of Lalitpur and Jhansi districts will get affected due to the Ken Betwa link project. No credible answer from NWDA or Union Government was given.

3. **Investments in Rajghat and Matatila Dam to be waste?** The investment made by UP Government on Rajghat Dam will become waste due to implementation of this link project, said UP officials. The response from the Government of India officials was interesting, in that they did not deny that these investments would become a waste. In fact they said that by the time the four projects of MP in upstream of Rajghat in Betwa basin (namely the Berari, Kesari, Richhhan and Neemkheda dams proposed under KBL project) are constructed, the existing projects like Rajghat and Matatila would recover their cost. Hence, what is clear from this is that current and

projected benefits from both Rajghat and Matatila Dams will be seriously affected as a result of KBL. It is shocking to learn that, thus, KBL will in fact affect the benefits from a number of existing projects in Ken and Betwa basin and this revelation comes out for the first time. This is not even mentioned in the Ken Betwa project reports. This is another reason in fact to show how unviable the KBL proposal is.

4. **Hydropower generation from Rajghat and Matatila Power Houses would be affected:** When Government of UP officials raised this issue and also mode of compensation, NWDA Director General also agreed that the loss of power from the existing projects would have to be assessed. The official minutes further says, "Secretary (WR) stated that he fully agrees with the apprehension of the Government of UP." The Secretary, MWR, however, said that the computation of losses can be done when the project is approved. This again makes it clear that KBL is leading to loss of benefits from existing projects. Officials agree that such losses are neither computed, nor taken into consideration while preparing the Feasibility Report. This further reinforces the conclusion that the KBL is not feasible.

5. **Water sharing on Ken River:** There is no provision for extra water for UP in addition to present available water utilised by UP, whereas extra water should be distributed equally between UP and MP.

As per NWDA, of the 6066 MCM water available at GGD, Committed Water is 3248 MCM: 2221 MCM for u/s use in MP and 1027 MCM for Ken Canal in UP. Remaining water (2818 MCM), should be equally distributed between the two states as per UP claim. MP does not agree. MP is not ready to supply extra water other than 37 TMC, which is as per 1972 agreement.

UP has demanded that more water to the extent of about 12.5 TMC should be given for domestic and industrial uses in addition to 60 TMC for uses downstream of Daudan Dam and that computation of regeneration from upper areas of Daudhan Dam should be more realistic. MP disagreed and differences on this remain to be resolved as on other issues.

While the official minutes of the January 2005 meeting are indeed revealing a lot of new issues and they also reinforce the non-viability of the link proposals, some of which are highlighted above, what is equally clear is that neither the state governments, nor the Union Govt. is trying to take the people of the region and nation into confidence about the wide spread implications of the River Link proposals. That is the fundamental characteristic of planning and decision-making of India's water resources development, unfortunately.

Basic Unresolved Issues

From the above analysis of available information on the proposed Ken Betwa Link Project, it is clear that the case for the proposal is fundamentally weak due to the following reasons, among others:

- **Permanent backwardness for GGD upstream areas?** It seems the areas upstream of the proposed GGD are slated to remain permanently backward. Significant portion of these areas are tribal areas. Very little provision of water has been made for these areas. And Feasibility Report suggests (p 17) an agreement between UP and MP that MP will not use more water in these areas than what has been allocated for it in the water balance. These areas that have not used developed water resources in the past are destined to remain backward forever, it seems. And the people of these areas do not even seem to know about this.
- **Problems with the Proposal:** The whole proposal is based on basic fallacies of surplus and deficit basins based on manipulated water balance calculations as shown above. If looked at closely, Ken basin does not have any surplus water and Betwa basin has many unexplored local options.
- **Need:** The proposal so far does not show convincing reasons about the need for the project.
- **Desirability:** Looking at the huge impacts that the project is likely to have in both basins, more destructive than beneficial, it is doubt if the project would prove desirable if all the costs are honestly added.
- **Decision-Making Process:** The local people have had no role in the decision-making process of the project. In fact, as is evident from the Feasibility Report and the NCAER survey they do not even know what this project is about and what

for. Nor do they need such a project, as is evident in answer to another question from NCAER.

- **Options:** It is clear from the above analysis that existing infrastructure in the two basins is not used to its optimum levels. There is huge scope for achieving more irrigation, water supply, power and flood management benefits from existing infrastructure. Moreover, there is also big scope for local options in terms of rainwater harvesting, groundwater recharge, groundwater use, watershed management, better cropping pattern and technique and so on. In fact such options would be the best way to increase the employment and growth potential in the two basins.
- **Costs:** The full costs of the project are not even known. When lower cost options are available, then why go for such project with huge costs and questionable and unsustainable benefits?
- **Viability:** Is the project and claimed benefits viable? It is clear from the water balance study given above that the project is not even viable in terms of availability of water or other resources are concerned.
- **Social and Environmental Impact Assessment:** It is clear from the above analysis that basic social and environmental impacts of the project are not even known. Even the land and forests to go under submergence for five of the six dams proposed and the canal system are not even known. Also the outstanding social and environmental issues from the dams constructed in the two basins show that there is little credibility that the authorities have either the will or the capacity to address these issues. If they want to show that rehabilitation is possible, they must show it by resettling the people already displaced and if they want to show that environmental issues can be addressed, the same must be shown by first addressing the outstanding environmental issues of existing projects in the two basins.
- **Drought and Floods:** One of the main justifications that has been forwarded for pushing river link projects has been to show how drought and flood problems can be solved by

Other Projects in the Region

S. No.	Project	Year	River	State	Type	Ht, m	Gross Res. Capacity, MCM	Purpose	Max. Discharge Capacity of Spillways	Catchment Area, sq. Km.
1	Beuchore	1964	Aralker	MP	TE	20	7.64	I	47	
2	Rajpur	1926	Arjun Nallah	UP	TE	13	6.62	I	193	
3	Barwa	1964	Barwa	UP	TE	20	39.49	I	535	
4	Barwar	1923	Baura	UP	TE	21	33.78	I	238	
5	Bilanadi	1973	Bilanadi	MP	TE/PG	32	63.69	I	448	
6	Burhanalla	1967	Burha	MP	TE	15	8.66	I	97	
7	Chandia	1927	Chandia Nala	MP	PG	25	5.86	I	506	
8	Kotra Khamba	1915	Hagninadi	UP	TE	18	3.82	I	65	
9	Halali Reser.	1976	Hallai	MP	TE	30	252.8	I	1189	
10	Jaiwanti	1929	Jaiwanti	UP	TE	15	9.43	I	44	
11	Salarpur	1960	Karipa	UP	TE	11	4.02	I		
12	Daroli Tank	1963	Karkra Str.	MP	TE	17	4.87	I	215	2.26
13	Khandeha	1929	Khandeha	UP	TE	14	2.69	I	128	
14	Khaprar Dam		Khaprar	UP			3.5	I,D		27.70
15	Tejgarh Tank	1959	Lamti Str.	MP	TE	26	6.79	I	285	
16	Aunjhar	1931	Local Str.	UP	TE	22	4.35	I	172	
17	Kabrai Lake	1956	Maingaria	UP	TE	15	13.22	I	340	
18	Magarpur	1920	Mind	UP	TE	16	2.46	I	173	
19	Motinala Tank	1957	Moti Str.	MP	TE	16	2.902	I	173	
20	Balmiki Ohen Sarovar Dam	1962	Ohen	UP	TE	24	38.37	I	1415	
21	Dongri	1986	Pahuj	UP				I,D		141.5
22	Pahuj Reser.	8th Pl	Pahuj	UP			15.0	I,D		
23	Devindra Nagar	1969	Sekra Str.	MP	TE	18	5.68	I	77	4.65
24	Mola	1929	Son Str.	MP	TE	22	19.96	I	615	2.26
25	Rampur Kalyan	1925	Stream	UP	TE	13	1.85	I	110	
26	Karahi	1973	Sunehi	MP	TE	18	3.82	I	360	
27	Khapatia	1919	Thota	UP	TE	16	6.03	I	188	
28	Burha	1962	Tr- Berihadi	MP	TE	16	8.31	I	297	

transferring water from flood prone basin to a drought affected one. In this case, though, both basins experience the droughts and floods simultaneously. Both basins have similar topographical and hydrological features. Hence, there is no justification in pushing this project in that respect either.

Rainfall and Groundwater Potential of Bundelkhand Region and Nearby Area

SN	District	Area, sq. Km.	Rainfall, mm	Annaul Replenishable Groundwater, MCM	% Groundwater Development (1990)	Forest Cover (%)	Cultiavated area (Ha)
1.	Banda, CSN	7624	1024	1426	12	11	498047
2.	Chhatarpur	8687	1083	1036	24.40	10	387000
3.	Damoh	7306	1115	829	6.24	36.3	38.6%
4.	Datia	2038	900	313	22.18	9	
5.	Hamirpur, M	7166	794	1229	12	5	508890
6.	Jalaun	4565	776	1239	10	6	341818
7.	Jhansi	5024	822	896	93	6	365512
8.	Lalitpur	5039	822	669	36	13	218995
9.	Panna	7135	1248	812	5.80	34	317410
10.	Sagar	10252	1279	1434	10.66	28	
11.	Tikamgarh	5048	1045	818	33.67	13	331586

As shown above the groundwater potential of many of districts have not been utilised properly. In the districts of Panna, Damoh and Sagar in MP the groundwater utilisation is only 5.8 per cent, 6.24 per cent and 10.66 per cent, while the Replenishable Groundwater Resources are 812 MCM, 829 MCM and 1434 MCM respectively. Similarly in Jalaun, Hamirpur and Banda districts in UP the groundwater development is only 10 per cent, 12 per cent and 12 per cent, while the Replenishable Groundwater Resources are 1239 MCM, 1229 MCM and 1426 MCM respectively. Rest of the districts have same position and no district has used more than 40 per cent of the potential, except Jhansi district, which has developed 93 per cent of potential.

Gross Irrigation Utilisation from Various Minor Irrigation Sources in Ken and Betwa Basins in 1993-94

Madhya Pradesh

Ha

	District	Dugwells	Shallow Tubewells	Deep Tubewells	Surface Flow Schemes	Surface Lift Schemes	Total
1.	Chhatarpur	96760	455	286	5533	11524	114558
2.	Damoh	17878	8074	747	9057	22531	58287
3.	Datia	50263	—	—	14	50	50327
4.	Guna	42166	11666	2635	4201	22962	83630
5.	Panna	11289	596	340	10518	15061	37804
6.	Raisen	26749	28753	880	2687	23493	82562
7.	Sagar	52117	4014	601	3293	29389	89414
8.	Satna	72031	13190	9939	4401	17426	116987
9.	Tikamgarh	114763	2355	1910	3760	5400	128188
10.	Vidisha	24714	14370	650	1783	155762	197279
	Total	**508730**	**83473**	**17988**	**45247**	**303598**	**959036**

Source: Report on Census of Minor Irrigation Scheme 1993-94, Ministry of Water Resources, Governmentt of India, 2001, Vol. II-VI.

Net Sown Area and Irrigated Area in Madhya Pradesh 1993-94

Ha

SN	District	Net Area Sown	M & M Schemes	Groundwater	Others	Total (C+D+E)	F as % of B
	A	B	C	D	E	F	G
1.	Chhatarpur	355071	1655	52265	24096	**78016**	**21.97**
2.	Damoh	295214	1562	15220	34726	**51508**	**17.45**
3.	Datia	131203	1775	18271	6684	**26730**	**20.37**
4.	Guna	569419	5042	27669	33234	**65945**	**11.58**
5.	Panna	233234	804	6619	24348	**31771**	**13.62**
6.	Raisen	414234	41570	27938	23723	**93231**	**22.51**
7.	Sagar	531981	4122	24645	35492	**64259**	**12.08**
8.	Satna	352085	1414	30262	22665	**54341**	**15.43**
9.	Tikamgarh	243365	2440	54736	15730	**72906**	**29.96**
10.	Vidisha	496078	26936	21468	36264	**84668**	**17.07**
	Total	**3621884**	**87320**	**279093**	**256962**	**623375**	**17.21**

Source: Report on Census of Minor Irrigation Scheme 1993-94, Ministry of Water Resources, Government of India, 2001, Vol. VI.

Uttar Pradesh

Ha

S. No.	District	Dugwells	Shallow Tubewells	Deep Tubewells	Surface Flow Schemes	Surface Lift Schemes	Total
1.	Banda	27431	32238	22790	228	1947	84634
2.	Hamirpur	5809	31559	5459	—	21	42848
3.	Jalaun	8908	16681	22609	197	936	49331
4.	Jhansi	95278	8483	220	98	311	104390
5.	Lalitpur	48337	721	—	—	—	49058*
6.	Mahoba	31438	518	—	—	—	31956
	Total	**217201**	**90200**	**51078**	**523**	**3215**	**362217**

Source: Report on Census of Minor Irrigation Scheme 1993-94, Ministry of Water Resources, Government of India, 2001, Vol. II-VI.

Note: In 1996, the Minor irrigated area Lalitpur had gone upto 71077 Ha, according to Progress Report of Minor Irrigation Programme (Jhansi Division) Oct. 1996.

Net Sown Area and Irrigated Area in Uttar Pradesh 1993-94

Ha

S. No.	District A	Net Area Sown B	M & M Schemes C	Groundwater D	Others E	Total (C+D+E) F	F as % of B G
1.	Banda	521964	6428	3253	114326	**124007**	**23.76**
2.	Hamirpur	310885	29564	43909	20787	**94260**	**30.32**
3.	Jalaun	336104	110468	37578	7359	**155405**	**46.24**
4.	Jhansi	292360	50482	52852	47173	**150507**	**51.48**
5.	Lalitpur	219004	37099	49067	0	**86166**	**39.34**
6.	Mahoba	191746	1569	670	58811	**61050**	**31.84**
	Total	1872063	235610	187329	248456	**671395**	**35.86**

Source: Report on Census of Minor Irrigation Scheme 1993-94, Ministry of Water Resources, Government of India, 2001, Vol. VI.

Note: The Net sown area figure for Lalitpur has been taken from Sankhyikiya Patrika, State Planning Inst UP, 1994 for Lalitpur for the year 1991-2 as the figure given in Minor irrigation Census (39095 Ha, which was less than the total irrigated area in the above table) was apparently wrong.

References

Central Board of Irrigation and Power (1979) Register of Water Resources Projects in India, New Delhi.

Central Board of Irrigation and Power (1987) Large Dams in India, New Delhi.

Central Water Commission (1996), Storages in River Basins of India, New Delhi.

Dams, Rivers & People, brought out by SANDRP, New Delhi, various dates, www.narmada.org/sandrp

Dr Bhartendu Prakash *et al.*, (1998) Problems and Potentials of Bundelkhand with Special Reference to Water Resource Base, CRDT, IIT, Delhi.

Ministry of Water Resources, Government of India (March 2001) Report of the Census of Minor Irrigation Schemes, New Delhi.

National Council for Applied Economic Research, Agro Economic, Socio Economic and Environmental Survey of Six Link Projects.

National Water Development Agency (1996) Feasibility Study of the Ken Betwa Link Project, downloaded from Government of India's official website on Task Force Inter Linking of Rivers, http://riverlinks.nic.in, last accessed on October 8, 2005.

Research Foundation for Science, Technology and Ecology, "Ken Betwa Link: A People's Assessment," New Delhi.

Uttar Pradesh Irrigation Department website, http://upgov.up.nic.in/irrigation/dp.htm accessed on various dates.

This Chapter has been adapted from the original paper "Analysis of Ken-Betwa River Link Proposal," Published by South Asia Network on Dams, Rivers and People (SANDRP), New Delhi, April 2005.

7

Critique of Ken-Betwa Feasibility Report: Resettlement, Rehabilitation, and Environmental Aspects

MEDHA PATKAR
PAYAL PAREKH

Introduction

Acting on a directive from the Supreme Court the government created a Task Force on River Linking in December 2002 (which has now been disbanded) to investigate the possibility of interlinking 16 peninsular rivers and 14 Himalayan rivers for the stated purposes of flood control, drought prevention, and irrigation. The Ken-Betwa link proposes to transfer 'surplus' waters of the Ken River to the Betwa Basin for primarily irrigation purposes benefiting Uttar Pradesh and Madhya Pradesh.

In this chapter we outline the problems associated with resettlement/rehabilitation and the environment due to large infrastructural projects. We then propose a framework for minimising displacement and adverse environmental impact and end by evaluating the Feasibility Report of the Ken-Betwa link within this context.

Large scale development projects result in large scale displacement requiring resettlement and rehabilitation of communities, often down-trodden and marginalised. There are also many adverse environmental impacts associated with such projects. The Ken-Betwa link will be no different.

Resettlement and Rehabilitation

Large scale projects, whether in water, power, mining or infrastructure sector, are often controversial, in part due to the displacement of communities and livelihood that they cause. This in turn leads to a conflict between those that are pushing the project forward, such as the government, international lending institutions and contractors, and those that are opposing the project; they are usually the down-trodden and marginalised communities with a rich natural resource-base that provides the capital for the project. These people, in most cases, *adivasis*, *dalits*, farmers, or labourers, are not considered and treated as 'investors', but are rather expected to sacrifice for the 'common good.'

What leads to such a situation? The mainstream paradigm of development does not value communities dependent on resource-based economies that use, possess or occupy and also protect natural resources more than others do, as it is the key to their sustainable livelihood. Therefore, social and environmental impacts of large projects are not adequately accounted for. Additionally, in today's capitalistic society, only when a tangible economic value can be placed on something, is it considered. But how does one place a price on the loss of culture, original forest or access to other resources such as fish?

The families and communities, therefore, demand at the very least a fair and just compensation for the loss of habitat, land, and other sources of livelihood. Resettlement and rehabilitation policies and laws have evolved as a result of people's struggles. There is now a national rehabilitation policy since 2003, not without flaws and weaknesses. There are also state level acts (as in Madhya Pradesh and Maharashtra) and sectoral policies, such as the National Thermal Power Corporation or Ministry of Mining guidelines. There is also a constitutional mandate to protect scheduled castes, scheduled tribes and others right to life. These communities that should be the first beneficiaries of any 'development' project are not properly rehabilitated and instead moved forcibly and against their will. In such a situation, communities are forced to struggle for the right to their land and livelihood.

It is estimated that large scale dams have displaced up to 40 million people in India alone. Due to the sad state of land records, many people, especially *dalits* and *adivasis* are not accounted for. They are then viewed as 'encroachers' on their own land, often denied compensation and rehabilitation. Thus, only 20-25 per cent of dam oustees have been rehabilitated in India. From Bhakhra oustees to

Tehri oustees, there has been no proper rehabilitation with regards to land, resources and compensation.

In the case of dams, it does not only occur within the reservoir nor at the beginning of a project. Adverse effects, such as degradation of land and, thus, livelihood and displacement reverberate to canal areas, as well as areas downstream of the dam during later phases of construction. In upstream areas, communities are deprived of the right to water, as is evident from the scores of dams in Maharashtra and elsewhere throughout India. People, especially in *adivasi* areas, get affected due to environmental mitigation strategies such as compensatory afforestation or the creation of national parks and sanctuaries. Communities are also often not informed in adequate time of their displacement or adequately compensated resulting in lives that are worse-off after the construction of the 'development' project affecting all spheres of life: social, cultural and economic.

In a country as densely populated as ours it is not possible to have a large scale infrastructure project without displacing communities. What should a framework look like that is just and respects the rights and needs of communities displaced by the Interlinking Rivers project? It is now accepted that development projects should be planned with the participation and consent of would-be affected people, which was approved in the Panchayat Extension to Schedule Areas Act (PESA, 1996) in the case of scheduled tribes. After a thorough consultation, if displacement is inevitable, it should be kept to a minimum. This may mean that the design and scope of the project is altered to cause minimal damage in terms of irreversible environmental degradation and displacement. And in the event of such displacement, all entitlements of the affected people must be recognised and protected, or their losses must be adequately compensated.

Rehabilitation must occur before the impacts of displacement are felt. Additionally compensation should not be a 'one-size fits all' package. Rather compensation must be appropriate for the community that is being displaced. The cut-off date for resettlement should be extended until the completion of the project. As even the World Bank stipulates, the rehabilitation must result in a situation in which oustees live in a better state than prior to displacement, not just economically, but also socially. Thus, provisions must be made to ensure their livelihood and access to natural resources, as well as preservation of community structure and culture.

Even with a sound plan for resettlement and rehabilitation can we ensure that it is carried out and enforced? The track record is not very good – the 1976 Maharashtra Rehabilitation Act that stipulates land for land compensation has not been followed in a single case of displacement due to a dam. PESA, which stipulates that displaced communities must be informed and consent twice – once prior to land acquisition and again prior to resettlement, is rarely enforced. If the resources and capacity is not there to ensure the implementation of a comprehensive resettlement and rehabilitation package, all we are left with are meaningless plans on paper. Thus, our focus must be on developing solutions to our water resource problems that minimise displacement and value the protection of natural resources.

Environmental Concerns

It has been accepted that there are negative environmental impacts associated with dams. Adverse environmental impacts due to dams include submergence of prime agricultural land, the destruction of various aquatic flora and fauna, submergence of forests, waterlogging and salination of soils and an increase in seismic tremors. A single project, the Sardar Sarovar is estimated to destroy at least 14,000 hectare of forest upon completion. In the case of the Narmada Sagar, over 40,000 hectare of forest land was destroyed and not compensated for. Waterlogging and salination occur after canals pass through soils that are ill-suited to absorb such large concentrations of water. By the late-1980s, approximately 17 per cent of irrigated land was saline. In the Chambal basin alone, almost 100,000 hectare are waterlogged.

Ninety per cent of large projects are sanctioned without the compliance of environmental regulations. If this scenario is to change, then each river ecosystem must be considered holistically and cannot simply be viewed for its potential water supply to humans. Comprehensive studies estimating the environmental impact of a project must be undertaken before clearances are granted. Projects that violate environmental laws such as the Environmental Protection Act (1986), Forest Protection Act (1980) and the Wildlife

Protection Act (1972) cannot be given the go-ahead. Environmental Impact Assessments are never done in our country to stop a project on environmental grounds or to evaluate how to mitigate or minimise the impacts of a project, but rather to find a way to push through the project despite environmental concerns. This undermines the efficacy of the EIA. During the cost-benefit analysis, short-term benefits must be weighed against long-term impacts. If the true costs of social and environmental impacts are factored in, many projects would not have a favourable benefit to cost ratio.

Review of Social and Environmental Concerns: Ken-Betwa Feasibility Report

Let us review the rehabilitation and resettlement plans and environmental evaluations as outlined in the Ken-Betwa Feasibility Report within the above context. Although, six dams are proposed for construction under the Ken-Betwa link, the Feasibility Report only discussed resettlement and rehabilitation and environmental impacts with regards of one of the dams, the Daudhan. It is very difficult to ascertain the feasibility of the link or to calculate the benefit to cost ratio when only one/sixth of the project is evaluated.

Social Impacts

Displacement of persons within the submergence area for only one dam is discussed. No estimates are given of those that will be displaced by the canal network or will experience degradation of their land in downstream areas. The report neither gives the name of the villages that will be submerged nor the exact number of people to be submerged. Rather, the Feasibility Report states, 'about 10 numbers. of villages having about 900 families with a population of about 8550 shall be affected.' Once the engineers have calculated the submergence area, one would expect that the government would carry out a detailed survey identifying the houses to be submerged and the exact number of inhabitants to be displaced. It is necessary for the government to also record the land owned or toiled on (in the case of landless labourers) to be able to adequately compensate land, as well as livelihood. Such information is necessary to adequately assess what costs are involved with regards to resettlement and rehabilitation.

The Feasibility Report does state that culturable land that is being submerged should be compensated within the command area, but the report does not actually state whether such land is available or how it will be acquired. If the land is acquired forcibly, it invariably leads to conflict and legal disputes. While an estimate is given of the costs associated with obtaining land, this is clearly an underestimate if this land is to be obtained from irrigated areas. It is also important that the land is contiguous. Additionally, there is no discussion about compensation of forest that *adivasis* often use for grazing or collecting forest produce.

There is absolutely no discussion in the Feasibility Report concerning livelihoods of those displaced. Presumably, most to be displaced are agriculturalists, thus, appropriate land compensation and access to forest is paramount to ensuring livelihood. The report does discuss labourers for the construction of the dam coming from the displaced population. But this work is temporary and may very well be seasonal. It cannot and should not be construed as a replacement for their previous livelihoods, but rather as a supplement. There is no discussion regarding livelihood replacement for other categories of workers, such as artisans or shopkeepers.

The report discusses resettlement of the displaced families into seven villages with approximately 1500 persons each. It is important the community structure and culture is not destroyed due to displacement. Often resettlement occurs in a region that is geographically different (hills *versus* plains) and culturally very different that may not be acceptable to the oustees.

The Feasibility Report does not mention how or whether the government plans to get prior and informed consent from those that will be displaced by the project. The report has also not laid out a process for adjustment of the project plan to minimise displacement.

Additionally, it is essential that the Ken-Betwa project is compared to other proposals that are able to fulfill the same objectives. Alternative solutions may very well exist that have fewer adverse social impacts.

Environmental Impacts

In order for such a project to receive clearance from the Ministry of Environment, a number of studies need to be carried out beforehand. There is no discussion of downstream impacts. Much of the work is still pending – for example, data from the soil surveys is not available, making it difficult to assess the likelihood of waterlogging and salination. The region is already prone to endemic malaria, but no vector analysis has been done. It would seem that the construction of a large reservoir (standing water) will only make conditions more suitable for mosquito breeding.

The magnitude of forest that is to be submerged is not small and is an area that is full of wildlife. 6400 ha out of a total of 8650 ha to be submerged is forest land, with 45 km^2 of land within Panna National Park to be flooded. This is approximately 9 per cent of the land set aside for preservation within the park. Has the Ministry of Environment and Forests given clearance? What about the animals living in the park? The available area for the animals is significantly reduced. Will there be compensatory afforestation? How will it be carried out?

The report has also suggested that pisciculture will be possible due to the creation of the reservoir. But, it is extremely difficult to maintain appropriate environmental conditions conducive for fish in an artificial environment. Is the government willing to provide resources to ensure that this is done?

The report states that there are no monuments in the region, but it is near to the famous temples of Khajuraho, suggesting that this is a region where excavations may need to be done in the future. The report does not discuss this possibility.

The Feasibility Report's assessment of the environmental impacts of the Ken-Betwa link is extremely superficial, making it difficult to calculate the true environmental cost of the project. Many questions arise regarding waterlogging and salination of soils, possible increase in malaria, submergence of forest land and downstream impacts. Again, the report only discusses the impact of one of six dams proposed for the link.

Conclusion

Due to the lack of detail in the Feasibility Report, the true environmental, resettlement and rehabilitation costs of the project cannot be calculated, making it impossible to assess whether the project is 'feasible' or not. The sloppiness with which the Ken-Betwa Feasibility Report has assessed displacement and environmental concern gives one little faith that the government will implement such a mammoth project in a socially just manner. Instead, it causes one to shudder at the negative impacts the down-trodden and marginalised communities and the environment will have to contend with.

8

Facilitating Water Transfers – Preparing and Marketing of Water Transfer Reports

A.D. MOHILE

1. Introduction and Scope

The Background

In the last 23 years, the National Water Development Agency (NWDA) has prepared a number of water balance, permissibility and feasibility reports for different links included in the National Water Perspective. These reports have been discussed in various official for a such as the Technical Advisory Committee of NWDA, Governing Body of NWDA etc. However, these reports were not available to the people at large including knowledgeable persons in different discipline out side the government.

In the last few years, for various reasons, all such proposals have attracted a large amount of public attention. This has resulted in considerable criticism of the proposals, and at least some of the criticism appeared uninformed. In this background, the decision to make the Feasibility Report of the Ken-Betwa link project (NWDA March, 1995), available to all, is a welcome move.

While the comments on the National Perspective Plan, and on the proposals of individual links have been far and wide, even after weeding out some comments as extreme and irrelevant, the main comments can be classified in the following groups:

1. Comments which indicate that the necessary legal and political instruments which will allow the implementation of the proposal have not been discussed
2. The comments, which express that the alternatives available to the water transfer proposal, for meeting the same broad economic and social objectives, are not discussed adequately.

Such alternatives would include: (a) alternative to water development and irrigation, to meet the objectives, as also (b) alternatives available within the realm of water development through in basin development, watershed management, demand management etc.

3. Comments, which indicate that enough information, has either not been collected or has not been included in the report.
4. Comments, which express that, the report is mainly an engineering document, in which the economic, social, legal, environmental and ecological and political issues are not addressed adequately.
5. Comments, which indicate that the financial feasibility of the project in terms of investments and returns from the investments is not addressed.
6. Comments regarding the detailing of the project proposals.

The Purpose

The purpose of this paper is to clarify the broad concepts, and to give suggestions for improving both the process and contents of the water transfer reports. Such improvements can, more effectively, address the various criticisms and comments expressed on the water resources development process in India, in general, and the contents of the Feasibility Reports, in particular. All these suggestions are preliminary and need to be discussed widely.

The author had been involved in the various proposals of NWDA. Thus, the paper is, in part, self-critical. The inherent strengths of the studies are not questioned; the manner of presentation is questioned. The changed readership and the involvement of larger group in a consultative appraisal of the proposals require such improvements.

2. The Concepts

Classifying Water Transfers

Transferring water, in space, in time, or in both the domains, to change the natural availability into what suits the human needs is the essence of all water development. In that sense water transfers are involved in all water development proposals, from small to large.

The National Perspective Plan, however, involves much larger transfers in term of water quantities and lengths. The water transfers can be classified as follows:

a) **Inter-basin water transfers** – This involves transfer of water from one basin to another. The concept of a basin involves the consideration of the basin as a unitary whole, in which action at any place can have repercussion else where in terms of water availability and water development possibility. The basin concept involves the concept of a "common terminus" beyond which the interest ceases to exist, since the action which the basin would not have any significant effects beyond the terminus. Generally, the mouth to the sea is considered as common terminus, but at times the definition of the terminus be require a change for further discussion, see para 5.2 in Mohile (2003).

   While many of the links in the National Perspective involve inter-basin water transfers, most of the northern links such as Ken-Betwa are within the large Ganga-Brahmaputra-Meghna (GBM) basin and are not inter-basin water transfer proposals.

b) **Inter-basin water transfers to non co-basin States** – This is a class of inter-basin water transfers, which allow the water to be used by political units beyond those involved in the basin. This classification would be of particular importance from the legal point of view.

c) **Long Distance Water Transfers (LDWT)** – Even comparatively smaller projects in small coastal basins would involve inter-basin water transfers from one basin to another, often within the same States. Meanings existing conventional irrigation projects have canals, which cross over low water divides to enter into another basin. These short distance inter-basin transfers have been implemented without special problems. The term LDWT represents much larger and long-distance transfers, which have special problems, even if they are not inter-basin.

d) **Interlinking of Rivers (ILR)** – This term in common use in India for the last few years appears to be a loose and undefined concept, which does not capture the essence of the proposals, made. The author prefers to call the present proposals in the National Perspective as LDWTs.

The Conceptual Plan, the Links, and the Projects

The National Perspective Plan is a broad plan, rather conceptual, its main concept is to distribute the water resources available to India in a more equitable fashion, after considering the need for equity on one hand and constraints of economic feasibility and social/environmental acceptability on the other. This is not a single plan of action to be implemented either in totality or to be differed.

Within this broad plan, there are a number of proposals for links. Some links are independent and can be implemented on their own without consideration of the other links. On the other hand, some links need to be considered together as a chain of inter-dependent links, all of which need to be considered together. The feasibility of the individual links in such a chain can not be separated out without meeting gross assumptions about the allocation of cost and benefits beyond the artificial frame of reference, which segregates one link from the other.

In techno-economic terms, a project would consist of all action necessary for obtaining the desired benefits. Thus, the water transfer links, alongwith the necessary storage structures on the upstream, the actions for primary, secondary and tertiary water distribution, field water application, drainage of water, conjunctive use of ground water etc. would form the project. The techno-economic feasibility has to consider the whole project and not the links in isolation.

These concepts are brought out because the reports of feasibility have to be about projects which may involve LDWTs, one or more, but not about the links alone.

3. The Broad Economic Objectives and the Alternatives

The Indian water development in general and irrigation development in particular has undergone slow shifts in the objectives. In the early-50s when the development started, food self-sufficiency was perhaps considered as an objective in itself, to avoid the large imports of food. Somewhat later, the Planning Commission prescribed the economic criteria of maximising the economic benefits over costs. The need for incorporating the distributional objectives was brought in later, by prescribing a smaller cut-off ratio (B/C of 1.0 against the normal of 1.5) in regard to tribal areas. The further improvements in terms of incorporating the shadow prices of in-puts and out-puts etc. were recommended in the report of the Nitin Desai Committee (Central Water Commission, 1983); but this have not been implemented so far.

Thus, national economic efficiency has been the main consideration in deciding the economic feasibility of the project. Since the entire agricultural planning continued on the basis of achieving and maintaining the self sufficiency in food, and restrictive trade practices in allowing food imports (except to meet shortages) was followed, food prices higher than the world market could be used to justify developmental projects. On the other hand the costs also were worked out at the prevalent national market prices. In the endemic unemployment situation, the shadow price of labour would be much lower than the minimum wages, but these shadow prices were not considered in the economic analysis. The implementation of the Nitin Desai Committee report could have allowed a reduction of economic costs by allowing the use of these and other shadow prices of inputs, as also a reduction in the value of the benefit by deciding the food prices on the basis of the world market.

The situation has changed considerably from the 1980s to the present time. The food self-sufficiency is already established. In the WTO regime, the world market is already affecting the Indian market. We need to closely examine if the food self-sufficiency is still to be pursued as a separate objective. Alternately, this can be pursued indirectly through trade barrier to be worked out under WTO. As a third alternative, we may allow large-scale international trade in food, or in other words, we may import virtual water instead of importing water from long distance.

According to the author, in our pre-occupation with national economic efficiency, we did not explicitly address the larger issues of distributive efficiency of benefits and connected issues of livelyhood and rural poverty. In India, a rapid urbanisation accompanied by considerable population growth has been projected. The National Commission (Ministry of Water Resources, 1999) projects a population of 1346 million under the low growth scenario and 1581 million under the high growth scenario for 2050. Due to large urbanisation, the rural population would be around 610 to 700 million. Much of this rural population would have derived its main income from the land (agricultural or animal husbandry), although, there could be some supplementation of income through seasonal migration etc. In particular, in the low rainfall areas w·th considerable population, the income from rain-fed agricultural would be so low that considerable part of the rural population would be below the poverty line. Drinking water to such areas through long distance water transfers would increase the economic carrying capacity of the area and reduce the incidence of poverty. The International Commission on Irrigation and Drainage (ICID) forcefully brought this forward in the recent studies during study of Sabarmati basin under the Country Policy Support Programme (CPSP).

The International Water Management Institute, Sri Lanka (IWMI) alongwith the International Food Policy Research Institute, Washington (IFPRI) are working on an economic-water use model, WATER – SIM. This model models the international agricultural economy including the international trade in agriculture. It thus, follows the third alternative mentioned above. The model is driven by market mechanism, and demand and supplies decide prices, productions and trade. The un-published preliminarily result reviewed by the author indicate that by 2025 India would be meeting its food demand by importing about 27 million tonnes of food, 15 million tonnes of milk and about 1.5 million tonnes of meat and poultry. Under such scenario, the need for water and the need for water development and water transfers may be relatively less. As stated, these results appear preliminary. But the main question is – if this is a globally optimal situation, can India accept this as a "*fait accompli*". And if this happens, how would the Indian rural population derive its income through the reduced food production?

Perhaps the NWDA studies need to be based on a more detail study of these basic aspects. These studies would meet the objections of the type 2 (a) mentioned in para I. From the work done so far, the author is of the opinion that India may have to use trade barriers to keep its rural population engaged in irrigated agriculture so that they derive enough income and are above the poverty line. If the economic models are right, such a strategy may be slightly inefficient from the national economic efficiency, but this would have to be accepted. The long distance water transfer proposals should have their main objective not as economic efficiency but as a measure for providing livelyhoods and alleviation of poverty. In individual project report also, it would not be sufficient to simply estimate the benefits. The distribution of benefits amongst various economic strata, and the effect of these benefits on livelihoods and poverty need to be presented after proper analysis.

Alternatives to the LDWT projects within the water sectors need to be dealt in each project report on a case-by-case basis, in detail. The Ken-Betwa report (NWDA 1995) hardly deals with the alternative of in basin development through demand management, water shade management, changing cropping patterns to suit water availability etc. The cropping pattern as proposed have very large paddy areas and these do not seem to reflect the scarcity of the value of the water imported through LDWT. This has already been criticised by Alagh (2005). The alternatives of more efficient but costlier distribution and application (such as drips, sprinklers, lined canals, piped distribution etc.) need to be discussed and devaluated.

4. The Information Base for Planning

Data About Rural Incomes

As already stated in section 3, much larger database about rural incomes and livelihoods and about the likely distribution of benefits would have to be integrated in the project planning.

Information About Soils

The sustained productivity of land, under irrigated agriculture is important for appraising the LDWT project. In the Ken-Betwa project report not much information about the soils in the proposed irrigated area is available. Information given in its section 5.2.2 seems to pertain to the whole basin and not the irrigated area. The information as presented gives an impression that 44 per cent of the area to be irrigated has plateau soil which are unfit to crops, etc. A considerably larger information base for the soils in the area to be serviced, (as differentiated from the basin or the zone), conversion of this information in crop suitably information, and the use of the information in deciding input packages, cropping patterns and sustainable outputs is necessary. Thus, cropping patterns and intensities as well as inputs and outputs would vary within the command, as per the main soil types. This is a detail planning exercise. Although, the full exercise can be in the Detailed Project Report (DPR), considerable information needs to be included in the feasibility reports.

5. Improving the Social and Environmental Contents of the Project

General

The criticism that project is more of an engineering report appears somewhat valid. In the historical context, water development reports were prepared by engineering organisations, for appraisal by other engineers and economists and the reports necessarily were engineering documents. While the action proposed in the LDWT reports would continue to be predominately engineering based the appraisal would now be done by a much larger community which would be inter-disciplinary. The information base, the detailing and style of writing, therefore may, have to change.

The Social Issues

As an example the Ken-Betwa report contains little information of sociological relevance. The population densities, their composition and differences with regional or national average in terms of income, literacy, general issues etc. has not been discussed. Some of these issues could be relevant to the detailing of the project and these may not have received attention. For example, the author was informed that his region of India has the largest density of cattle. Animal husbandry and dairy seems to be an important sector in its future development, but this has not been reflected. On the other hand, a planner can argue that the large livestock is undesirable, and in future larger mechanisation of farm operations and negative growth of cattle needs to be planned. Such a line also has not been taken. Such issues need to be discussed, a need to be reflected in the planning of the crops, inputs, etc. The generally cutup nature of the topography, and the small fragmented land holdings, (as differentiated from large contiguous tracks of fertile plans in the main Gangetic valley) is a feature of the region. In another context, the author had expressed that in such situations; *Shejpali* may be a better option for irrigation management than *Warabandi*. The planning of the tertiary distribution, however, has not been discussed at all.

The Environmental and Ecological Issues

There has been much criticism about the likely environmental impacts of the LDWT projects and their affect on the ecology of the region. Therefore, dealing with these in the LDWT reports becomes even more important than in a conventional water development projects. This would require a much better appreciation and internalisation of the concerns of the eco-scientists (as differentiated from the un-informed criticism of activist without scientific knowledge). For example the relevant chapter of the Ken-Betwa report deals with aquatic resources in terms of production of reservoir fisheries. This is pisciculture, a productive process for livelihood and economic development similar to agriculture. The concern of the ecologists is about the life cycle of the fish native to the river, and about the likely changes in their life due to the water development. These could occur because of the obstructions, reductions in flows or quality problems. Again, the real concern is about the whole ecologic chain, and is not limited to the fish. These issues have not been dealt at all. Similarly, the project reports mentions that there are large mammals like tigers in the area. Will forest submergence reduce their territories by submerging corridors? This has reportedly happened in reservoirs in the Western Ghats in Maharasstra. Such issues have been ignored.

A particular concern in regard to LDWTs is the possibilities that flora and fauna that is alien to a river may be imported into it through the transfer from the donor river. This could then cause a sudden change in the eco-system. Although, such possibilities appear far fetched in the Indian context, these need to be discussed by comparing the flora and fauna in the two rivers. This has not been done.

6. The Financial Aspects

A conventional water development project in India is prepared by a State Government, which fully finances it through the plan funds, and recovers revenues as per its tariffs. Normally, such projects show a negative return on investment. Although, the Union policies require recovery of atleast the O&M cost, no one really bothers, at the appraisal stage, about the negative returns. For LDWT, the situation could be different. Most proposals involve more than one State. The implementation cost could be met either by the Union or by the States. The revenues should; normally be collected and kept by whoever invests. The tariffs could be higher than the normal state projects because these are non-conventional and costly projects. If Union is investing it needs to perhaps ensure that its policies in regard to cost recovery are incorporated. Private financing is a possibility, which is being talked about. Another possibility being mentioned is to obtain loans from public or from financing institutions including international institutes or banks. If such alternatives are followed, recoveries could become important and some principal of cost recovery somewhat similar to those for private hydro development could become relevant. Such issues need to be considered and discussed. The financial analysis as presented in the Ken-Betwa report appears to be without any purpose and makes no one wiser.

7. The Detailing of the Proposals

The proposals would, as stated, require many more details with regard to economic objectives, distribution of benefits, poverty and livelihoods, soils, crop planning, irrigation planning and application, etc.

The conjunctive use of rainwater, surface waters and ground waters is another area, which requires more detailing. The donor basins could be "open", that is, it may not be using all the water. The use of rainwater could, therefore, enhance utilisation. The recipient basin could be "closed" and rainwater use may affect other uses, requiring regulation The studies need to cover both surface and ground waters, and need to consider the interactions in planning. For example, both percolation tanks and reuse of regenerated surface flows through diversions or small tanks may be possible. The resources could be increased in the process.

8. Concluding Remarks

The improvements indicated in this paper would require:

a) Finalisation though wider discussions.

b) Deciding what is essential for feasibility, and what can be done in the detailed project report.

c) Adopting more pragmatic approach for completed feasibility studies through essential supplementation, on one hand and proceeding with the process of reaching broad agreements, on the basis of available information, on the other hand.

References

Alagh, Y.K. (2005). "On Interlinking of Rivers," *The Indian Express*, New Delhi, April 6.

Central Water Commission, (1983). *Report of the Committee to Review the Existing Criteria for Working out the Benefit Cost Ratio for Irrigation Projects*, Government of India.

International Commission of Irrigation and Drainage, (2004). *Report on Water Policy Issues: India Country Study-CPSP*, ICID New Delhi, August.

Mohile, A.D. (2002). "Theme Paper on Interbasin Transfer of Water in India- Prospects and Problems," *Water Management Forum, Institute of Engineers* (India).

Ministry of Water Resources, (1999). *Report of the National Commission for Integrated Water Resources Development, Vol. 1*, Government of India.

National Water Development Agency, (1995). *The Feasibility Report of the Ken- Betwa Link Project.*

Annexures

Annexure 1

"Feasibility Report of Ken-Betwa Link Project"
National Water Development Agency

EXECUTIVE SUMMARY

The Ken-Betwa link project envisages diversion of surplus waters of Ken basin to water deficit Betwa basin. The quantity of water proposed to be diverted from Ken basin, after considering in basin demands and downstream commitments earmarked for providing irrigation in Madhya Pradesh and Uttar Pradesh, is 1020 Mm^3. This link canal will provide irrigation to water short areas of upper Betwa basin of Madhya Pradesh by way of substitution and also to enroute areas of Madhya Pradesh and Uttar Pradesh. The command envisaged in the earlier proposed Ken Multipurpose Project (KMPP) by Madhya Pradesh State Government is also to be irrigated from this project. Apart from drinking water facility and enroute irrigation of 47000 ha in Chhatarpur and Tikamgarh districts of Madhya Pradesh and Hamirpur and Jhansi districts of Uttar Pradesh, provision for downstream commitments of 1375 Mm^3 for MP and 850 Mm^3 of water for UP has also been kept.

A dam is proposed on river Ken at Daudhan 2.5 km. upstream of existing Gangau weir. The 75 per cent dependable yield of Ken upto Daudhan site has been assessed as 6188 Mm^3. The net water availability at dam site after accounting all the upstream requirements is obtained as 3291 Mm^3. The downstream commitments from Ken at Daudhan are 2225 Mm^3. Out of which, 850 Mm^3 is provided to Uttar Pradesh and 1375 Mm^3 to Madhya Pradesh as per Interstate agreement of 1981 on river Ken. This 1375 Mm^3 will be utilised to irrigate the entire command of KMPP. The surplus water for diversion at Daudhan is 1020 Mm^3. Out of which, 659 Mm^3 is transferred to Betwa river upstream of Parichha weir and 312 Mm^3 is utilised in the enroute command.

The dam proposed at Daudhan is an earthen dam with a side channel spillway with FRL of 287 m. and gross storage capacity of 2775 Mm^3. The maximum height of the dam is 73.80 m. The total length of the side channel spillway is 326 m. 15 gates of 18 m. x 18 m. size are provided to pass the design flood of 45104 cumecs. Two power houses, one (Power House No.1) at the foot of the dam and other (Power House No.2) at the end of 2 km. long tunnel are proposed. The installed capacities of Power House No.1 and 2 are 3 x 20 MW and 2 x 6 MW respectively. The Power House No.1 will be a pumped storage scheme and the pondage between the existing Gangau weir and Daudhan dam will be utilised to store water releases from the Power House. This storage will be pumped back by the Francis type reversible turbine of this Power House during off peak periods.

The total length of the link canal from Ken at Daudhan to its outfall point is 231.45 km. including a 2 km. long tunnel in its head reaches. The design discharge of the link canal at its head is 72 cumecs. The canal is designed as a trapezoidal section with rounded corners and lined for the entire length. The full supply depth and bed width of the canal at its head are 3.56 m. and 12 m. respectively with 1 in 10,000 bed slope. The side slopes of the canal are 1.5 (H): 1 (V). The canal crosses several rivers/streams. The link canal after traversing about 230.00 km. outfalls in an existing reservoir namely Barwa Sagar from where the diverted water will join Betwa river through a natural stream in the upstream of Parichha weir.

The Betwa command comprises of four projects namely Barari, Neemkheda, Richhan and Kesari in the upper reaches of the Betwa basin. An area of 1.27 lakh ha in the Raisen and Vidisha districts of Madhya Pradesh will be benefitted by utilising 659 Mm^3 of water annually from this link by way of substitution. This link is also to provide annual irrigation to 47000 ha area enroute in the drought prone Chhatarpur and Tikamgarh districts of Madhya Pradesh and Hamirpur and Jhansi districts of Uttar Pradesh. The water use for this command is 312 Mm^3. Besides, an area of 3.23 lakh ha annually under earlier proposed KMPP command (termed as Ken command in this report) in the Chhatarpur and Panna districts of Madhya Pradesh will be benefited. The water use in this command is 1375 Mm^3, which will be met from releases of Daudhan dam into Ken river through Power House No.1.

The link will also provide 11.75 Mm^3 water for drinking water supply to the enroute villages of Chhatarpur and Tikamgarh districts of Madhya Pradesh and Hamirpur and Jhansi districts of Uttar Pradesh.

Detailed surveys and investigations of the headworks and the main canal have been carried out for establishing the feasibility of the project. The preliminary designs and preparation of cost estimates have been done on the basis of these surveys and investigations.

Auxiliary and incidental benefits that will occur by the commissioning of this scheme are of immense value to the people of the area. Ground water will get supplemented due to increased use of surface water for irrigation. The scarcity of drinking water in these dry areas will be mitigated considerably. The afforestation programme could be implemented on canal banks resulting in environmental improvement. The communication system will improve because of canal roads and CD works raising marketing opportunities. The formation of the reservoirs will help tourism development, fish and aquaculture, bird sanctuaries etc. Besides, there will be an increase in employment opportunities during the construction of project.

Interlinking river development programme, particularly those that include the development of one or more dams and reservoirs, can produce far-reaching changes in the environment of the region. Environmental effects, or changes resulting from project development, may occur upstream, on site, downstream or along the link project as well as changes in population distribution and land use in the immediate reservoir area as well as other zones due to increased development for irrigated agriculture, industry or other purposes (fisheries, resettlement). Adequately planned such big projects provide major economic benefits. However, as with many other developments, there may be direct and indirect adverse environmental and social effects (costs) as well as beneficial environmental impacts secondary to the initial project purposes. With careful planning, adverse effects can be minimised or mitigated, and secondary beneficial effects enhanced. Net benefits from the link project often will increase when it become regional development project that integrate irrigation, power generation and municipal water supply with catchment area management, conservation, tourism, fisheries and rural development. Thus, environmental and socio-economic analysis is one of the ways in which resource development project can be analysed, to identify and minimise potential resource conflicts/adverse environmental effects, thus enhancing overall project viability. It was felt necessary to carry out a study to access the likely effects on various aspects such as agro-economic, socio-economic and environmental impacts. National Council of Applied Economic Research (NCAER), New Delhi was engaged by NWDA to carry out the benchmark survey of the socio- and agro-economic conditions prevailing in the region of the link project.

The total cost of the link project has been estimated as Rs. 1988.74 crore (1994-95 price level) comprising Rs. 991.07 crore for Main Project (Unit-I: Head works Rs. 367.92 crore, Unit-II: Canals Rs. 572.44 crore and Unit-III: Power Rs. 50.72 crore). The cost of canal system is Rs. 554.11 crore for Ken command (i.e. KMPP) and the estimated cost of four projects is Rs. 443.55 crore for Betwa command. The schedule of construction of the link project is planned for a period of 9 years including pre-construction year. The annual benefits accrued from the project are estimated to be in the order of Rs. 449.79 crore (1994-95 price level). The rates at April 1994-95 price level have been worked out based on the Unified Schedule of Rates (USR), April 1991 of Water Resources Department, Govt. of Madhya Pradesh after considering escalation @ 10 per cent per annum.

The Benefit-Cost ratio for the Ken-Betwa link project as a whole has been computed based on the corresponding annual costs and annual benefits from this project, which comes to 1.87. The internal rate of return (IRR) of the project works out to 13.00 per cent.

SALIENT FEATURES

1.0	Name of the Project	Ken-Betwa Link Project	
2.0	General		
2.1	River basin(s)		
	(a) Name	Ken and Betwa river basins	
	(b) Located in states	Madhya Pradesh and Uttar Pradesh	
2.2	Name of		
	(a) Rivers	Ken and Betwa	
	(b) States	Madhya Pradesh and Uttar Pradesh	
	(c) Districts		
	(i) Reservoir	Panna and Chhatarpur of Madhya Pradesh	
	(ii) Head works	Chhatarpur	
	(iii) Command area	Chhatarpur and Tikamgarh of Madhya Pradesh and Jhansi and Hamirpur of Uttar Pradesh	
	(iv) Power houses	Chhatarpur	
	(d) Tehsils		
	(i) Reservoir	Bijawar of Chhatarpur district and Panna of Panna district	
	(ii) Head works	Bijawar of Chhatarpur district	
	(iii) Power house	Bijawar of Chhatarpur district	
	(iv) Command area	Chhatarpur and Nowgong of Chhatarpur district, Jatara and Tikamgarh of Tikamgarh district, and Mahoba of Hamirpur district and Mauranipur of Jhansi District	
	(e) Village near the head works	Daudhan	
2.3	Access to the project	Name	Distance from project site
	(a) Air port	Khajuraho	31 km (by road)
	(b) Railway station	Satna	110 km (by road)
	(c) Road	Chhatarpur-Panna State Highway	19 km (W.B.M.road)
2.4	Estimated life of the project	100 years	
3.0	**Type of the project**	Multipurpose	
3.1	Irrigation		
	(a) Culturable Command Area	i Enroute command 0.47 lakh ha ii Ken command 2.41 lakh ha iii Betwa command 1.02 lakh ha	

Contd...

...Contd...

	(b) Annual irrigation	i Enroute command 0.47 lakh ha
		ii Ken command 3.23 lakh ha
		iii Betwa command 1.27 lakh ha
3.2	Power (Generation) (a) Installed capacity (MW)	i 3x20=60 MW in PH-1 (one unit standby)
		ii 2x6=12 MW in PH-2
	(b) Firm power	25 MW at 60% load factor
		i) Firm – 18 Mkwh (Mu) I Annual energy (MKwh)
		ii) Total – 191.67 Mkwh (Mu)
	(d) Cost of Power per Kw installed capacity	Rs. 7,044.83
4.0	Hydrology	
4.1	Catchment area upto dam site	19,534 Sq Km
4.2	Precipitation	
4.2.1	In the catchment area	Period from 1901 to 1983
	(a) Normal	1,174.07 mm (annual)
	(b) Maximum	1,662.10 mm (annual)
	I Minimum	660.00 mm (annual)
4.2.2	In the command area	Period from 1987 to 1991
	(a) Average annual	950 mm
	(b) Average monsoon	1,100 mm
	I Average non-monsoon	20 mm
4.3	Annual yield at Daudhan dam	6,188 Mm^3 at 75% dependability
4.4	Climatic data	
4.4.1	(a) Project area	
	(i) Temperature	44o C (Max.) & 6.7 oC (Min.)
	(ii) Humidity	95% in monsoon & 9% in non-monsoon
	(iii) Wind speed	16.1 km/hr (Max.) & 1.0 km/hr (Min.)
	(b) Command area	
	(i) Temperature	47.3 oC (Max.) & 4.0 oC (Min.)
	(ii) Humidity	95% (Max.) & 9% (Min.)
	(iii) Wind speed	4.2 km/hr (Av.)
4.5	Utilisation of water	
4.5.1	From Daudhan reservoir	
	(a) For direct irrigation	
	(i) In M.P	1,375.00 Mm^3

Contd...

...Contd...

	(ii) In U.P.	850.00 Mm^3		
	(b) For irrigation through link canal			
	(i) In Enroute command	312.00 Mm^3		
	(ii) In Betwa command	659.00 Mm^3		
	(c) For drinking purpose in the vicinity of link (MP & UP)	11.75 Mm^3		
	(d) Transmission losses 37.25	Mm^3		
	Total	3,245.00 Mm^3		
4.5.2	Committed utilisation in upstream projects	Major (Mm3)	Medium (Mm^3)	Minor (Mm^3)
	(i) Existing	-	42.31	162.54
	(ii) Ongoing	-	-	69.10
	(iii) Proposed	1,099.95	402.66	340.09
4.6	Design flood for spillway	45,104 cumec		
5.0	Reservoir/Weir			
A	Daudhan reservoir			
	(i) Water levels			
	(a) F.R.L.	287 m		
	(b) M.D.D.L.	268 m		
	(c) Dead storage level	238 m		
	(ii) Free board	4.0 m		
	(iii) Wave height	3.01 m (Max.)		
	(iv) Live storage	2,752.69 Mm^3		
	(v) Capacity at			
	(a) F.R.L.	2,775 Mm^3		
	(b) M.D.D.L.	1,390 Mm^3		
	(c) D.S.L.	22.31 Mm^3		
	(vi) Accumulated	50 years 100 years sedimentation after		
	(a) upto new zero elevation	25.0 Mm^3	75.0 Mm^3	
	(b) Above new zero elevation	199.73 Mm^3	311.76 Mm^3	
	(vii) Evaporation losses	Av. Annual	152.93 mm	
		Av. Non-monsoon	165.15 mm	
		Av. Monsoon	135.89 mm	
B	Weir at Gangau			
	(i) Water levels			
	(a) Storage level	234.90 m		

Contd...

...Contd...

	(b) River bed level	219.30 m
	(c) Live storage capacity	56.40 Mm^3
6.0	Submergence under Daudhan Dam	
6.1	Land and property submerged	
A	Area submerged at FRL	
	(i) Forest area	6,400 ha
	(ii) Culturable area	2,171 ha
	(iii) Unculturable area	79 ha
	Total	8,650 ha
B	Property affected at FRL	
	(a) Village affected	10 Nos.
	(a) Building/houses	750 Nos.
	(c) Wells	50 Nos.
	(d) Road/rails (Km)	No major roads and railway lines
6.2	Number of families affected	900 (approx.)
6.3	Number of persons affected	8,550 (approx.) (The detailed study on survey and investigation for rehabilitation & resettlement of families would be carried out and incorporated at the time of preparation of the detailed project report.)
7.0	Head works	
7.1	Dam	
7.1.1	Earth/rock fill dam	
	(a) Type of dam	Earthfill
	(b) Length of the dam at top	
	(i) Right flank	1,224 m
	(ii) Left flank	244 m
	(c) Top width	8 m
	(d) Maximum height above(foundation level)	
	(i) Right flank	73.80 m
	(ii) Left flank	19.95 m
7.2	Masonry/concrete dam	
	(a) Type of dam	Concrete
	(b) Elevation at top	291.0 m
	(c) Elevation of deepest foundation	232.0 m
	(d) Length at top	247.0 m (considering both sides of spillway)

Contd...

...Contd...

	(e) Width at the top	8 m
	(f) Maximum height above foundation	55 m
7.3	Spillway	
	(a) Type of spillway	Side channel
	(b) M.W.L.	288 m
	(c) Length	326 m
	(d) Max. height above deepest foundation	28 m
	(e) Crest level	269 m
	(f) No. of gates	15 nos.
	(g) Type of gates	Radial
	(h) Size of gates	18 m x 18 m
	(i) Design discharge	44,000 cumecs
	(j) Tail water level	234.75 m
	(k) Type of energy dissipation	Ski-jump bucket type device
8.0	Canal system	
8.1	Canal system	
	(a) Purpose of the canal	Irrigation
	(b) Type of flow	Gravity flow
	(c) Type of lining	P.C.C.
	(d) Length of the main canal	231.45 km (including 2 km long tunnel)
	(e) FSL at head	259.0 m
	(f) FSL at end	220.62 m
	(g) Bed width at head	12 m
	(h) Side slope	1.5:1
	(i) Bed slope	1:10,000
	(j) Discharge capacity	72 cumecs upto 134 km , 62 cumecs from 134 km to 195 km & 57 cumecs from 195 km to terminal point at Barwa Sagar
	(k) Total number of canal structures on main canal	59
(l)	Total assumed head losses across the structure	6.725 m
	(m) Gross command area	51,500 ha
	(n)Culturable command area	47,000 ha
	(o) Annual irrigation	47,000 ha Enroute and 1.27 lakh Ha. In Betwa command

Contd...

...Contd...

8.2	Branch canals		
	(a) Number	28 Nos.	
	(b) Total length	140 km	
9.0	Power (Components)		
9.1	Surge shaft		
	(a) Type	Simple	
	(b) Size	23.5 m dia	
	(c) Height above Foundation level	32.55 m	
	(d) Elevation at top	292.552 m	
	(e) Capacity	3,92,869 Mm^3	
	(f) Surge level	Max. –291.052 m, Min. –282.928 m	
9.2	Penstocks	for PH-1	for PH-2
	(a) Number	3	2
	(b) Diameter	4.0 m	3.7 m
	(c) Length	33.0 m	19.0 m
9.3	Power house		
	(a) Type Underground for	PH-1 and surface for PH-2	
	(b) Head	PH-1	PH-2
		Max.=52.25 m	Min.=33.25 m
		Max.=28.00 m	Min. = 9.00 m
	(c) Dimension	33m x 18 m	19m x 12.5 m
	(d) Installed capacity	3x20 MW	2x6 MW
	(e) Turbines		
	(i) Types	Francis vertical shaft (reversible)	Francis Verti cal shaft
	(ii) Number	3	2
	(iii) Capacity (H.P.)	26,820	8,046
	(f) Number of standby units	1x20 MW (Power House-1)	
	(g) Cost of power component	Rs. 5,071.86 lakh	
9.4	Tailrace	PH-1	PH-2
	(a) Type	a tail race tunnel, which outfalls in the pool between the existing Gangau weir and proposed Daudhan dam	a channel leading to K-B link canal
	(b) Water level	Max. - 235 m Max. - 259 m	Min. - 225 m Min. - 255.4 m

Contd...

...Contd...

	(c) Size of draft tube gate	2.5m x 8.2 m	2.1 m x 7.0 m
9.5	Cost (1994-95 price level) Cost of K-B link project		
	Unit – I Head Works	Rs. 36,791.63 lakh	
	Unit – II Canals (Link and Enroute command)	Rs. 57,243.77 lakh	
	Unit – III Power	Rs. 5,071.86 lakh	
	Sub-total	Rs.99,107.26 lakh	
	(b) Ken command: Canal system	Rs.55,411.32 lakh	
	(c) Betwa command: Upper Betwa projects	Rs.44,355.50 lakh	
	Total cost	Rs.1,98,874.08 lakh	
	(d) B.C. Ratio for the K-B link project as a whole	1.87	
	(e) Internal rate of return	13.00%	
	(f) Cost of irrigation	Rs. 40,000 per ha	
	(g) Cost of generation of power	54 Paise per Kwh	

Layout of the Proposed Ken-Betwa Link Canal

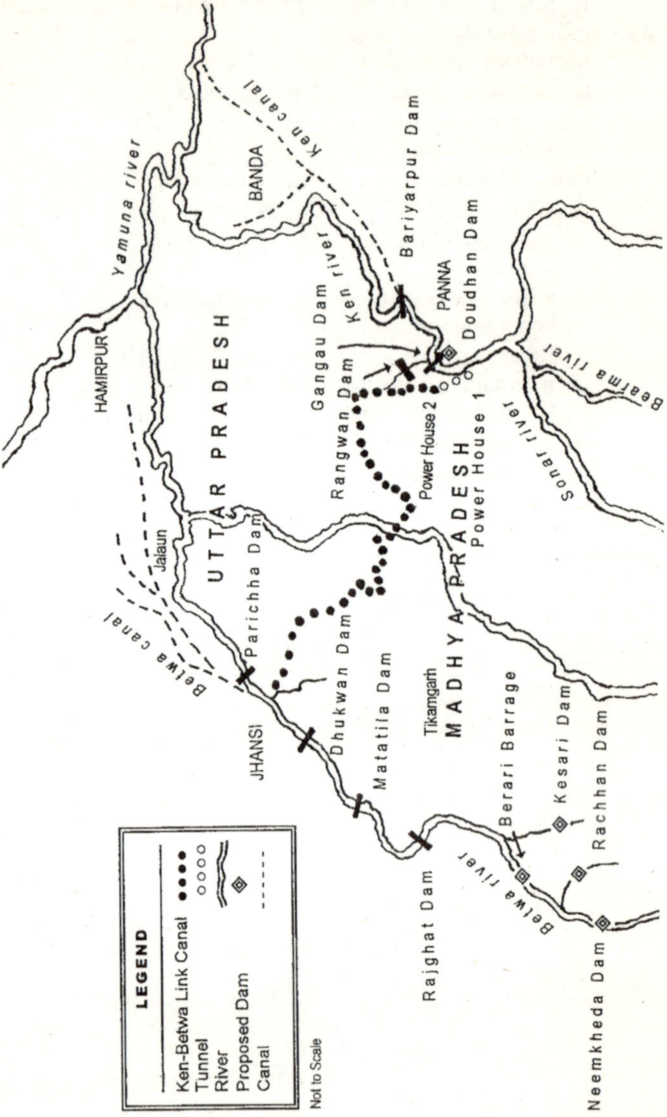

Annexure 2

National Civil Society Committee on Interlinking of Rivers in India

Given the size and significance of the proposed project it was realised that a national debate on the issue is an essential pre-requisite. Keeping this objective in mind, a meeting was organised in February 2003 at the India International Centre, New Delhi. Before the meeting, preparations were made meticulously. A detailed background document was prepared and circulated well in advance to the participants. The list of participants was drawn from people and institutions that expressed opinions about the project and have major stakes in the future in this project. Several Government officials, members of the Task Force, academics, civil society organisations, NGOs, movements, water resource management experts and organisations attended this meeting. In principle the participants agreed that a civil society commission or committee should be set up to hold a dialogue with the Government on this project. After this, several consultative meetings of small groups were held to explore the possibility of establishing such a committee, which could initiate a debate and carry on the dialogue focusing on the complexity of the project, its technical, financial economic implications, social and equity implications and likely ecological consequences. Thus, came into being the National Civil Society Committee on Interlinking of Rivers (NCSCILR.) which was established with 13 members on 20th October 2003.

The committee includes several individuals who represent the wide spectrum of public opinion about interlinking of rivers. However, it is essential to state here that it was not possible to include all the groups and individuals, working in the area, while setting up a civil society committee which is accepted by all. However, an attempt was made to include people who hold different opinions on the issue, which is essential to conduct the dialogue and get acceptance from all the sides.

The members of the Committee are:

Professor Y.K.Alagh (Chairman), Dr. Ashok Khosla (Executive Chairman), Dr. Biksham Gujja, Dr. S. Hashim, Professor Jayanta Bandyopadhyay, Professor Kanchan Chopra, Dr. Ramaswamy R. Iyer, Dr. E.A.S. Sarma, Mr. B.P. Singh, Mr. A.D. Mohile, Ms. Medha Patkar, Dr. Tushaar Shah, Mr. Ganesh Pangare and Mr. Nitin Desai.

The committee on a whole decided not to take any stand on the issue, but only facilitate the national dialogue in the coming years, which would provide inputs into the process of decision-making. The Committee agreed on the following broad objectives, subject to availability of resources.

- Generate public debate about the project.
- Facilitate and improve the sharing of information and ideas between civil society, Government, Task Force, NWDA and others.
- Bring in all the available knowledge and experiences from the past and build knowledge generated through independent studies.
- Advocate for appropriate policy framework.
- Strengthen the platform for civil society at large.

The Committee is of the view that the country needs to review the ILR project and come up with answers for some basic questions, before the project is implemented. Some of these questions are

- Is ILR the most cost-effective option for domestic water security in drought-prone areas in India?
- Is India's food security critically dependent on the interlinking project?
- Can the ILR project solve the flood situation in India?
- How to calculate water surplus and water deficit in rivers?
- Will inter-state and transboundary issues be addressed and resolved?
- What are the viable alternatives to it?

The establishment of the Civil Society dialogue on Interlinking of Rivers is somewhat unique in India in many ways. But in no way this is a perfect process. The process is intended to be consultative and inclusive as far as possible right from the beginning, especially as everybody expresses opinion, but there is not much interaction. To bring distinguished stalwarts from varied backgrounds having opposing viewpoints, on the same platform is an achievement in itself. Since the inception of the ILR project, the NCSCILR members had been urging the government and the ILR Task force to review and assess the project before implementation. The Government has kindly taken note of the views of the Committee and made the Ken-Betwa feasibility study available in the public domain. However, the task of the dialogue has hardly been easy. The process of committee formation has been long, and due to diverse views work has moved at a slower pace than expected. Also, the publicity of the issue and media attention has also made the process difficult. The unclear government stand on the issue for a long time has been an additional difficulty. Absence of reports and data in the public domain for a long time, has made any assessment difficult.

The work undertaken by the Committee includes the following:

- Collate, analyse and endorse work that has been done till now in specific links. Initiate fresh study to assess ground situation, wherever required.
- Review/initiate basin-wise water balance studies to understand surplus and deficit water situation, in the context of the ILR project.
- Review of overall foodgrain demand and the water necessary for it.
- Analyse ILR against the background of the established principles of integrated water resource management (IWRM).
- Review cost-benefit analysis of water resource development projects, in the context of ILR.
- Review of the existing guidelines and standards related to taking up of such large water resource development projects, (e.g. the Nitin Desai Committee Report) and Rehabilitation plans are drawn up for such mega projects.

- Impacts on environment and biodiversity in specific links.
- Environmental flow study in river basins, particularly in the context of ILR.
- Address policy, legal and Constitutional interventions necessary to address sound water resource development in India, especially in the context of the ILR.

This is a crucial moment for building India's water resources, conserving water and also shaping the country's ongoing environmental movement. There is a need for committed environmental movement on one hand, and the need to develop water resources on the other. But the essential question is to strike a fine balance between the two, ensuring sustainable development and hence this dialogue. The dialogue is trying to assess the proposed project in totality without stepping into the shoes of environmental activism or that of an active project proponent. It is expected to take place in a more evolved manner, based on new data, analysis and new ideas. It is a process where agreeing to disagree and still working together for finding solutions to a problem, hold the key. There is a long way to go before concluding that the ILR is the best available option to improve the water resource situation and control, floods and droughts in the country. It is important the Government is not only itself convinced about the potential of ILR, but is also able to convince the citizens and civil society, that it is indeed the best option.